Synthesis Lectures on Engineering, Science, and Technology

The focus of this series is general topics, and applications about, and for, engineers and scientists on a wide array of applications, methods and advances. Most titles cover subjects such as professional development, education, and study skills, as well as basic introductory undergraduate material and other topics appropriate for a broader and less technical audience.

Farzin Asadi

Applied Numerical Analysis with MATLAB®/Simulink®

For Engineers and Scientists

Farzin Asadi
Department of Electrical and Electronics Engineering
Maltepe University
Istanbul, Turkey

ISSN 2690-0300   ISSN 2690-0327   (electronic)
Synthesis Lectures on Engineering, Science, and Technology
ISBN 978-3-031-19368-2   ISBN 978-3-031-19366-8   (eBook)
https://doi.org/10.1007/978-3-031-19366-8

This Springer imprint is published by the registered company Springer Nature Switzerland AG
The registered company address is: Gewerbestrasse 11, 6330 Cham, Switzerland

Preface

Numerical analysis is the branch of mathematics that is used to find approximations to difficult problems such as finding the roots of nonlinear equations, integration involving complex expressions, and solving differential equations for which analytical solutions do not exist. It is applied to a wide variety of disciplines such as business, all fields of engineering, computer science, education, geology, meteorology, and others.

Many good books are written about the numerical analysis theories. In this book, we don't focus on the theories of numerical analysis. Instead we focus on solving numerical analysis problems with the aid of tools that MATLAB/Simulink provides for us. This book can accompany and complete any standard theoretical textbook on numerical analysis. The prerequisite for this book is a first course on numerical analysis, and no prior knowledge of MATLAB/Simulink is assumed.

This book is written primarily for students of engineering and science. However, it can be useful for anyone who wants to solve a numeric problem with MATLAB/Simulink. For instance, master or Ph.D. students can use the material presented here to solve the numerical problems of their thesis or papers. All of the material presented here can be covered in one semester (14 weeks) with 3 h per week.

This book is composed of 11 chapters. Here is a brief summary of each chapters:

Chapter 1 is an introduction to MATLAB. In this chapter, you will learn how to do basic calculations with MATLAB. Basic matrix operations, trigonometric, hyperbolic, logarithmic, exponential, and many other important functions are studied in this chapter.

Chapter 2 shows how MATLAB can be used for symbolic calculations. In this chapter, you will learn to calculate limits, derivatives, integrals (both definite and indefinite), partial fraction expansion, Laplace transform, Fourier transform, and Taylor series with MATLAB.

Chapter 3 shows how MATLAB can be used to calculate a definite integral (single, double or triple) or derivative of a function at a given point. Trapezoidal and Simpson rules for calculation of definite integrals are studied in this chapter.

Chapter 4 shows how MATLAB can be used for statistical calculations. In this chapter, you will learn how to calculate summation, average, variance, and standard deviation of a given data, how to generate random numbers, how to calculate factorial and combinatorial, and how to obtain cumulative distribution function for a normal distribution.

Chapter 5 shows how MATLAB can be used to obtain impulse and step responses of a linear differential equation. In this book, we use the linear system (or linear dynamical system) and linear differential equation terms interchangeably. By impulse and step responses, we mean the response that observed in the output when the input is Heaviside step function and Dirac impulse function, respectively.

Chapter 6 shows how differential equations can be solved with Simulink. Simulink is a software package which accompanies MATLAB and permits you to simulate dynamical systems (i.e., systems which can described by a linear or nonlinear differential equations) with the aid of many ready to use graphical blocks.

Chapter 7 shows how difference equations can be solved with Simulink. A difference equation is a relation between the differences of unknown function at one or more general values of the independent variable. Generally, a difference equation is obtained in an attempt to solve an ordinary differential equation by finite difference method.

Chapter 8 shows how curve-fitting problems can be solved with MATLAB's Curve Fitting Toolbox™. Curve fitting is the process of constructing a curve, or mathematical function, that has the best fit to a series of data points.

Chapter 9 shows how MATLAB can be used for drawing the graph of a given data. Different kind of graphs are studied in this chapter.

Chapter 10 shows how some of well-known numerical analysis algorithms can be implemented with MATLAB programming.

Chapter 11 shows how optimization problems can be solved with the aid of ready to use functions that Optimization Toolbox™ provides.

I hope that this book will be useful to the readers, and I welcome comments on the book.

Istanbul, Turkey   Farzin Asadi
                                                                                    farzinasadi@maltepe.edu.tr

Contents

Essential of MATLAB®

1

1.1 Introduction

MATLAB® (an abbreviation of MATrix LABoratory) is a programming platform designed specifically for engineers and scientists to analyze and design systems and products that transform our world. The heart of MATLAB is the MATLAB language, a matrix-based language allowing the most natural expression of computational mathematics. Keywords of MATLAB language are shown in Table 1.1.

Table 1.1 Keywords of MATLAB language

break	case	catch	classdef
continue	else	elseif	end
for	function	global	if
otherwise	parfor	persistent	return
spmd	switch	try	while

Millions of engineers and scientists worldwide use MATLAB for a range of applications, in industry and academia. MATLAB is both an analysis and design tool. This chapter introduces the MATLAB and its common daily uses for electrical engineers.

© The Author(s), under exclusive license to Springer Nature Switzerland AG 2023
F. Asadi, *Applied Numerical Analysis with MATLAB®/Simulink®*,
Synthesis Lectures on Engineering, Science, and Technology,
https://doi.org/10.1007/978-3-031-19366-8_1

1.2   MATLAB Environment

The MATLAB environment is shown in Fig. 1.1. It is composed of four parts. First part is a collection of icons that are required frequently. The second section (current folder browser) enables you to interactively manage files and folders in MATLAB. Use the current folder browser to view, create, open, move, and rename files and folders in the current folder. Use the icon shown in Fig. 1.2 to open the desired folder. The third section of MATLAB is command window. The MATLAB commands are entered here and their results are shown here. You need to press Enter key of your key board to run the commands written in command window. The fourth section of MATLAB (the workspace) contains variables that you create within or import into MATLAB from data files or other programs.

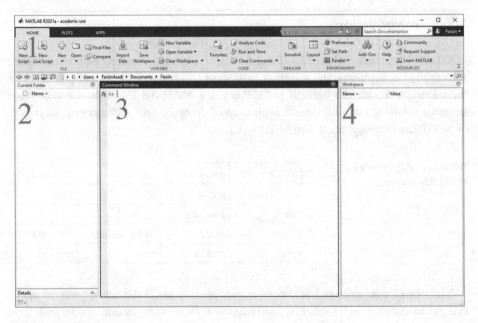

Fig. 1.1 MATLAB environment

Fig. 1.2 Brows for folder icon

The layout of MATLAB can be changed using the layout icon. This book uses the three column layout (Fig. 1.3).

Fig. 1.3 Layout icon

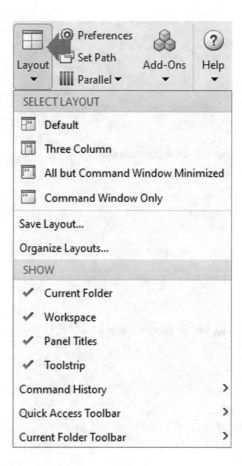

1.3   Basic Operation with MATLAB

Let's define a complex variable a with value of $1 + 2i$. This can be done with the aid of commands shown in Figs. 1.4, 1.5 or 1.6. In MATLAB, both of i and j can be used to enter the imaginary part of a complex number.

Fig. 1.4 Defining variable a

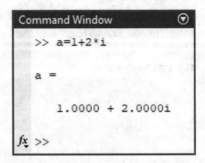

Fig. 1.5 Defining variable a

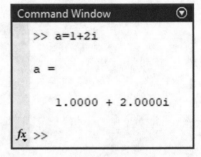

Fig. 1.6 Defining variable a

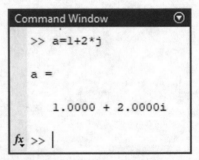

After running the commands of Figs. 1.4, 1.5 or 1.6, a new variable will be added to the workspace window (Fig. 1.7).

Fig. 1.7 Variable a is added to Workspace

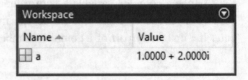

The result of commands will not be shown if you put a semicolon at the end of commands (Fig. 1.8).

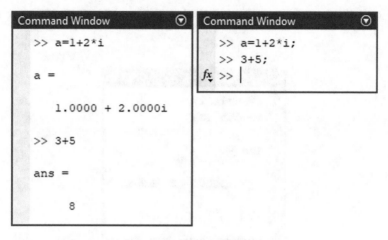

Fig. 1.8 Semicolon hides the results of commands

You can use the real and imag function to obtain the real and imaginary parts of a complex number (Fig. 1.9).

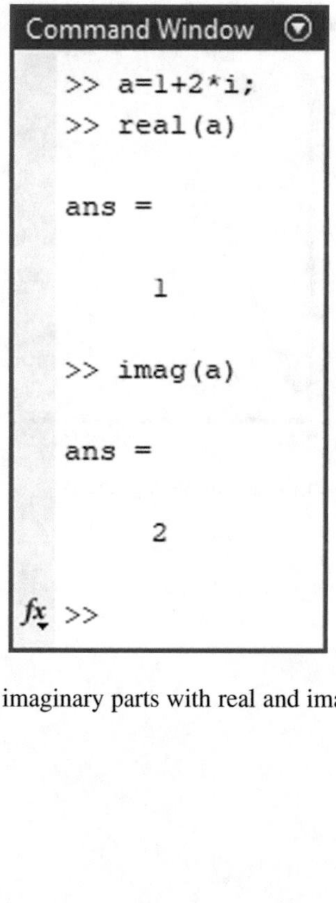

Fig. 1.9 Calculation of real and imaginary parts with real and imag commands

You can use the conj function to calculate the complex conjugate of a number (Fig. 1.10).

```
Command Window                    ⊙
   >> a=1+2i;
   >> conj(a)

   ans =

        1.0000 - 2.0000i

fx >> |
```

Fig. 1.10 Calculation of complex conjugate of a number with conj command

The point $a = 1 + 2i$ is shown in Fig. 1.11. The polar form this point is $\sqrt{1^2 + 2^2} e^{j \tan^{-1} \frac{2}{1}}$. Value of $\sqrt{1^2 + 2^2}$ and $\tan^{-1} \frac{2}{1}$ is calculated with the aid commands shown in Fig. 1.12. You can calculate the magnitude and phase of a complex number with the aid of commands shown in Fig. 1.13 as well.

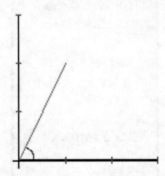

Fig. 1.11 Representation of $1 + 2i$ in the complex plane

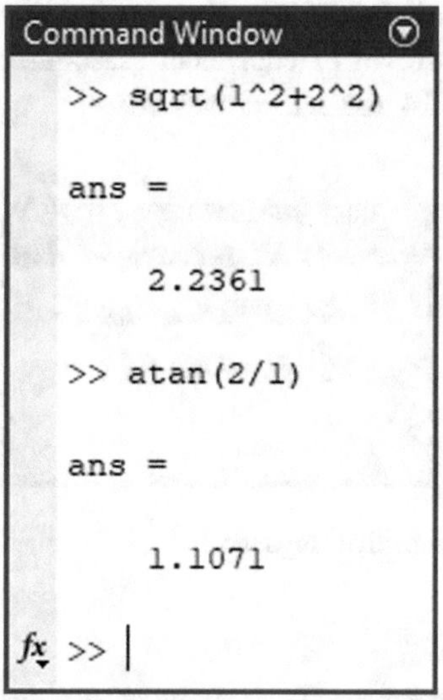

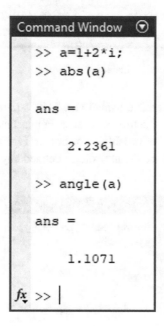

Fig. 1.12 Calculation of magnitude and angle of $1 + 2i$

Fig. 1.13 Calculation of magnitude and angle of $1 + 2i$

In computers, case sensitivity defines whether uppercase and lowercase letters are treated as distinct (case-sensitive) or equivalent (case-insensitive). MATLAB is a case sensitive language. Figure 1.14 and 1.15 proves this.

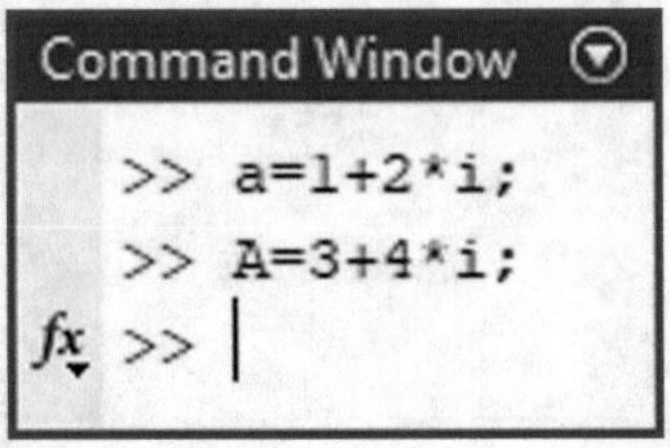

Fig. 1.14 MATLAB is a case sensitive language

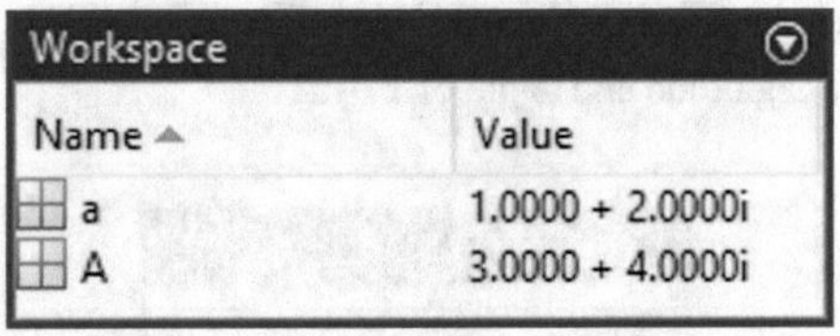

Fig. 1.15 A and A are two different variables

MATLAB has a default variable named ans. This variable is used to save the results of commands when you don't define a variable. For instance, in Fig. 1.16 the result of multiplication will be put in the variable c however in Fig. 1.17, ans will be used to save the result of multiplication since the user didn't defined any variable.

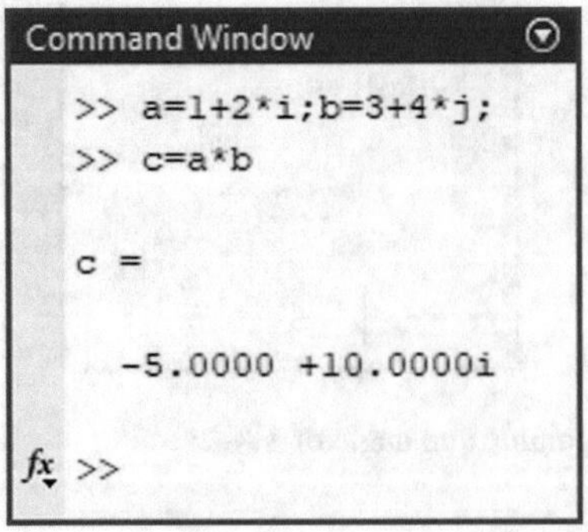

Fig. 1.16 Variable c equals to product of a and b

Fig. 1.17 Default variable ans
saves the calculation result

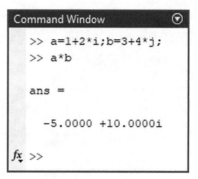

Commands in Fig. 1.18 do some basic operations on two complex numbers. The percent symbol (%) is used for indicating a comment line.

Fig. 1.18 Basic operations in
MATLAB

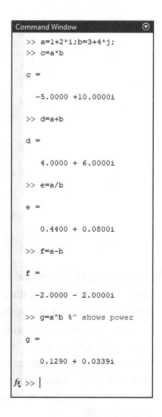

The reminder of division of two real numbers can be found using the mod or rem functions (Fig. 1.19).

Fig. 1.19 Rem and mod
commands

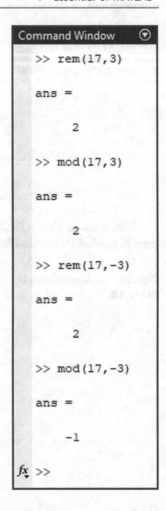

The concept of remainder after division is not uniquely defined, and the two functions
mod and rem each compute a different variation. The mod function produces a result that
is either zero or has the same sign as the divisor. The rem function produces a result that
is either zero or has the same sign as the dividend.

Another difference is the convention when the divisor is zero. The mod function
follows the convention that mod(a,0) returns a, whereas the rem function follows the
convention that rem(a,0) returns NaN. NaN stands for "Not a Number" and is used to
represents values that are not real or complex numbers. Expressions like 0/0 and inf/inf
result in NaN.

Both of mod and rem functions have their uses. For example, in signal processing, the
mod function is useful in the context of periodic signals because its output is periodic
(with period equal to the divisor).

1.4   Clearing the Screen and Variables

You can clear the command window using the clc command. Result of clc command is shown in Fig. 1.20 (Fig. 1.21).

Fig. 1.20 Clc commands
clears the command window

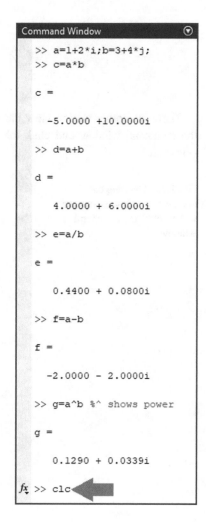

Fig. 1.21 Commands window
is cleared with clc command

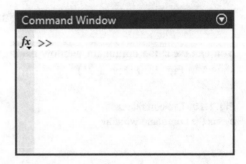

There is another way to clear the command window as well. You can right click on the command window and click the clear command window in the appeared window (Fig. 1.22).

Fig. 1.22 Clicking the clear
command window is another
way to clear the command
window

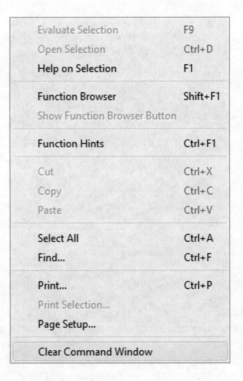

Clearing the command window does not affect the variables (Fig. 1.23). If you want to clear all the variables, you need to use the command shown in Figs. 1.24 or 1.25. If you want to clear a specific variable, then you need to write its name after the clear command. For instance, the command shown in Fig. 1.26, clears the variable g only. There is another way to remove the variable g as well. You can right click on it and click the delete from the appeared menu (Fig. 1.27).

Fig. 1.23 Clearing the command window has no effect on the variables

Workspace	
Name ▲	Value
a	1.0000 + 2.0000i
b	3.0000 + 4.0000i
c	-5.0000 + 10.0000i
d	4.0000 + 6.0000i
e	0.4400 + 0.0800i
f	-2.0000 - 2.0000i
g	0.1290 + 0.0339i

Fig. 1.24 Deleting all the workspace variables

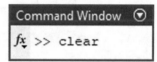

```
Command Window
fx >> clear
```

Fig. 1.25 Deleting all the workspace variables

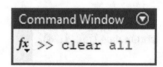

```
Command Window
fx >> clear all
```

Fig. 1.26 Deleting the variable g of Workspace

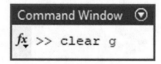

```
Command Window
fx >> clear g
```

Fig. 1.27 Deleting the
variable g of Workspace

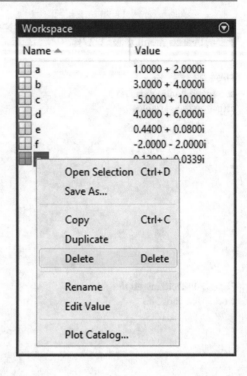

You can use the up and down arrows of the keyboard to access the previous entered
commands (Fig. 1.28).

Fig. 1.28 Accessing to the
previously entered commands

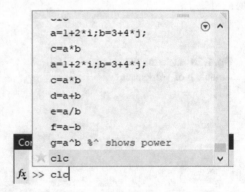

1.5 Basic Matrix Operations

Anything in MATLAB is a matrix. Even a number is considered as a 1×1 matrix. Assume that you want to enter $A = \begin{bmatrix} 1 & 2 & 3 \end{bmatrix}$, $B = \begin{bmatrix} 4 \\ 5 \\ 6 \end{bmatrix}$ and $C = \begin{bmatrix} 7 & 8 & 9 \\ 10 & 11 & 12+2i \\ 13 & 14 & 16 \end{bmatrix}$ to MATLAB. The commands shown in Figs. 1.29 or 1.30 do this job.

Fig. 1.29 Defining the matrices A, B and C

```
Command Window
>> A=[1 2 3];
>> B=[4;5;6];
>> C=[7 8 9;10 11 12+2*i;13 14 16];
fx >> |
```

Fig. 1.30 Defining the matrices A, B and C

```
Command Window
>> A=[1,2,3];
>> B=[4;5;6];
>> C=[7,8,9;10,11,12+2*i;13,14,16];
fx >> |
```

You can read the element on the i'th row and j'th column of matrix X, by using the command X(i, j). If matrix X is $1 \times n$, then you can use X(i) instead of X(1, i). In the same way, If matrix X is $n \times 1$, then you can use X(i) instead of X(i,1). Generally, the term vector is used for $1 \times n$ and $n \times 1$ matrices.

The commands shown in Fig. 1.31, gives some example for reading the elements of matrices defined in Figs. 1.29 and 1.30.

Fig. 1.31 Accessing the
elements of the matrices

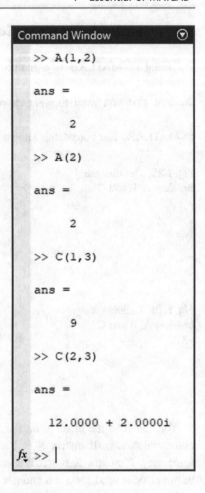

The commands shown in Fig. 1.32, changes the element in the second row and third column of matrix C to 12 + 12i.

```
Command Window                                                    ⊙

   >> C(2,3)=12+12*i;
   >> C

  C =

      7.0000 + 0.0000i    8.0000 + 0.0000i    9.0000 + 0.0000i
     10.0000 + 0.0000i   11.0000 + 0.0000i   12.0000 +12.0000i
     13.0000 + 0.0000i   14.0000 + 0.0000i   16.0000 + 0.0000i

  fx >> |
```

Fig. 1.32 Changing the element on the second row and third column of C

You can use the size command to read the number of rows and columns (Fig. 1.33).

Fig. 1.33 Calculation of
number of rows and columns
of a matrix

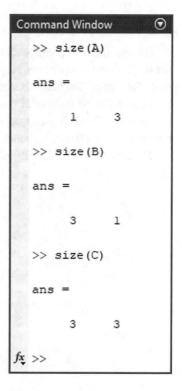

You can save the number of rows and number of columns in variables as well. The commands shown in Fig. 1.34 save the number of rows and columns of matrix A in RowA and ColumnB variables, respectively.

Fig. 1.34 Number of rows and columns of matrix A are saved in Row A and ColumnA, respectively

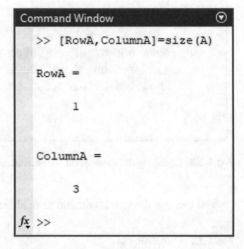

The multiplication, summation and subtraction of two matrices can be done with the aid of *, + and − operators, respectively. Note that the size of two matrices must be the same for summation and subtraction and the number of columns of the first matrix must be the same as the number of rows of the second matrix, otherwise the operation will not be done. MATLAB has elementwise operators as well. The element wise operator does the operation on the corresponding elements. For instance, in Fig. 1.35, [1 2 3].^[4 5 6] = [1^4 2^5 3^6] = [1 32 729] or [1 2 3].* [4 5 6] = [1 * 4 2 * 5 3 * 6] = [4 10 18].

Fig. 1.35 Elementwise
commands

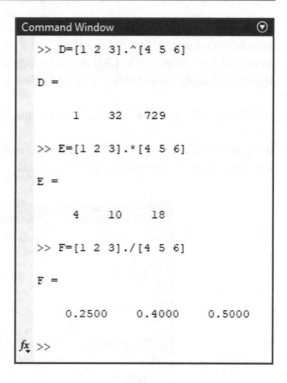

You can find the minimum and maximum of a vector by using the min and max
commands, respectively (Fig. 1.36).

Fig. 1.36 Calculation of
minimum and maximum of
vector A

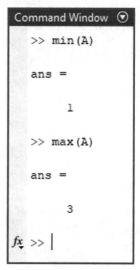

The min /max function can be applied to matrices as well. If the input of these functions is a m × n matrix, they return a n × 1 vector containing the minimum/maximum element from each column (Fig. 1.37). When the matrix is complex, the minimum/maximum is computed using the magnitude (Fig. 1.38). Remember that the magnitude of a complex number $a + bi = \sqrt{a^2 + b^2}$. In Fig. 1.38, the magnitude matrix is $\begin{bmatrix} 7 & 8 & 9 \\ 10 & 11 & 16.971 \\ 13 & 14 & 16 \end{bmatrix}$. So, the max function returns $12 + 12i$ as the maximum of the third column since $12 + 12i$ has the maximum magnitude in the third column.

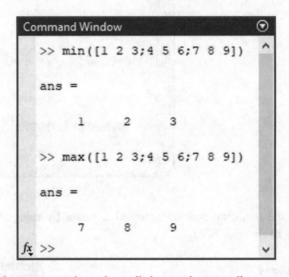

Fig. 1.37 Min and max commands can be applied to matrices as well

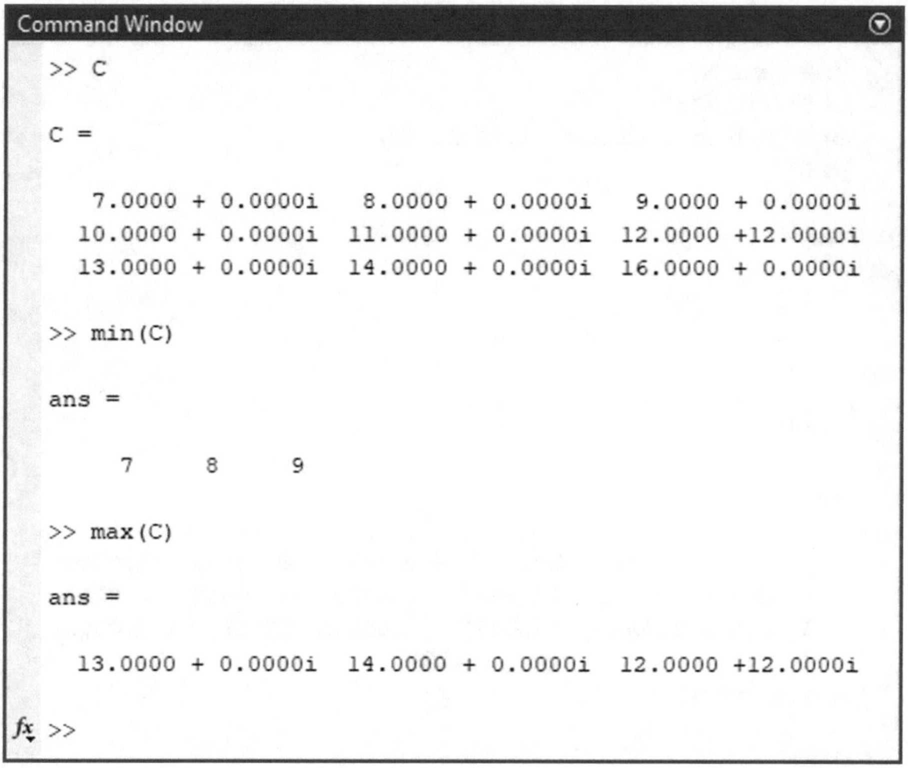

```
Command Window                                                    ⊙

   >> C

   C =

        7.0000 + 0.0000i    8.0000 + 0.0000i    9.0000 + 0.0000i
       10.0000 + 0.0000i   11.0000 + 0.0000i   12.0000 +12.0000i
       13.0000 + 0.0000i   14.0000 + 0.0000i   16.0000 + 0.0000i

   >> min(C)

   ans =

           7       8       9

   >> max(C)

   ans =

       13.0000 + 0.0000i   14.0000 + 0.0000i   12.0000 +12.0000i

  fx >>
```

Fig. 1.38 Calculation of minimum and maximum element of matrix C

The transpose of a matrix can be calculated using the single quote operator or transpose function (Fig. 1.39). If the matrix is real, then two methods produce the same results. However, when the matrix is complex, the results are not the same. The transpose function simply flips the matrix over its diagonal (i.e. switches the row and column indices of the matrix), however the single quote function calculates the complex conjugate of elements and then change the row and column indices.

```
Command Window                                                          ⊙

  >> A=[1 2 3];
  >> B=[4;5;6];
  >> C=[7 8 9;10 11 12+2*i;13 14 16];
  >> A'

ans =


      1
      2
      3


  >> C'

ans =

     7.0000 + 0.0000i   10.0000 + 0.0000i   13.0000 + 0.0000i
     8.0000 + 0.0000i   11.0000 + 0.0000i   14.0000 + 0.0000i
     9.0000 + 0.0000i   12.0000 - 2.0000i   16.0000 + 0.0000i

  >> transpose(C)

ans =

     7.0000 + 0.0000i   10.0000 + 0.0000i   13.0000 + 0.0000i
     8.0000 + 0.0000i   11.0000 + 0.0000i   14.0000 + 0.0000i
     9.0000 + 0.0000i   12.0000 + 2.0000i   16.0000 + 0.0000i

fx >> |
```

Fig. 1.39 Different methods of calculation of transpose of a matrix

You can use the det function to calculate the determinant of square matrix. The inverse of a matrix X can be calculated using the inv(X) or X^{-1} commands.

Assume that you want to ensure that the two matrices calculated in Fig. 1.40 are the same. To do this we need to calculate the difference between the two matrices. Figure 1.41 shows that the obtained results are the same.

```
Command Window                                                    ⊙

  >> det(C)

  ans =

    -3.0000 +12.0000i

  >> inv(C)

  ans =

    -2.3529 - 0.0784i    0.0392 + 0.1569i    1.3137 - 0.0784i
     2.1176 - 0.1961i    0.0980 + 0.3922i   -1.2157 - 0.1961i
     0.0588 + 0.2353i   -0.1176 - 0.4706i    0.0588 + 0.2353i

  >> C^-1

  ans =

    -2.3529 - 0.0784i    0.0392 + 0.1569i    1.3137 - 0.0784i
     2.1176 - 0.1961i    0.0980 + 0.3922i   -1.2157 - 0.1961i
     0.0588 + 0.2353i   -0.1176 - 0.4706i    0.0588 + 0.2353i

fx >>
```

Fig. 1.40 Calculation of inverse of matrix C

Fig. 1.41 Calculation of
difference between the C^{-1}
and inv(C) commands

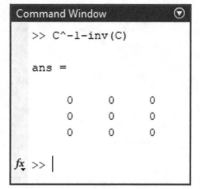

```
Command Window                    ⊙

  >> C^-1-inv(C)

  ans =

              0      0      0
              0      0      0
              0      0      0

fx >> |
```

Assume that you want to solve the $\begin{bmatrix} 7 & 8 & 9 \\ 10 & 11 & 12+2i \\ 13 & 14 & 16 \end{bmatrix} x = \begin{bmatrix} 4 \\ 5 \\ 6 \end{bmatrix}$. The two commands shown in Fig. 1.42 can be used to calculate the vector x.

Fig. 1.42 Calculation of vector x

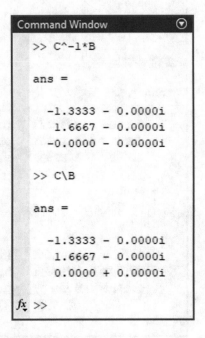

```
Command Window

>> C^-1*B

ans =

    -1.3333 - 0.0000i
     1.6667 - 0.0000i
    -0.0000 - 0.0000i

>> C\B

ans =

    -1.3333 - 0.0000i
     1.6667 - 0.0000i
     0.0000 + 0.0000i

fx >>
```

The Fig. 1.43 shows that the difference between the result of commands shown in Fig. 1.42. Note that 1e-15 mean 10^{-15}. So, the difference between the results is very small.

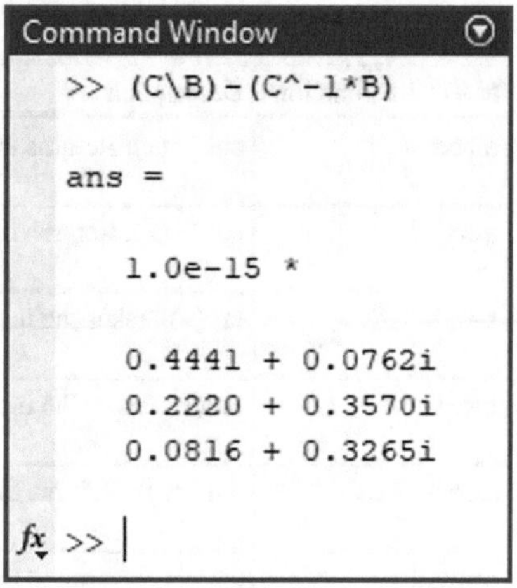

Fig. 1.43 Calculation of difference between the C\B and $C^{-1} * B$ commands

```
Command Window                          ⊙

   >> (C\B)-(C^-1*B)


   ans =


      1.0e-15 *


      0.4441 + 0.0762i
      0.2220 + 0.3570i
      0.0816 + 0.3265i

 fx >> |
```

1.6   Trigonometric Functions

Trigonometric functions of MATLAB are listed in Table 1.2. x can be a complex number as well.

Table 1.2 Trigonometric functions of MATLAB

MATLAB function	Description	Example
deg2rad	Takes an input in degrees and returns it equivalent in radians	≫deg2rad(90) ans = 1.5708
rad2deg	Takes an input in radians and returns it equivalent in degrees	≫rad2deg(pi) ans = 180
sin	$\sin(x)$. Takes the input in radians	≫sin(pi/2) ans = 1
sind	$\sin(x)$. Takes the input in degrees	≫sind(30) ans = 0.5000
asin	$\sin^{-1}(x)$. Returns the output in radians	≫asin(0.5) ans = 0.5236
asind	$\sin^{-1}(x)$. Returns the output in degrees	≫asind(0.5) ans = 30.0000
cos	$\cos(x)$. Takes the input in radians	≫cos(1) ans = 0.5403
cosd	$\cos(x)$. Takes the input in degrees	≫cosd(60) ans = 0.5000

(continued)

Table 1.2 (continued)

MATLAB function	Description	Example
acos	$\cos^{-1}(x)$. Returns the output in radians	≫acos(0.5) ans = 1.0472
acosd	$\cos^{-1}(x)$. Returns the output in degrees	≫acosd(0.5) ans = 60.0000
tan	$\tan(x)$. Takes the input in radians	≫tan(1) ans = 1.5574
tand	$\tan(x)$. Takes the input in degrees	≫tand(60) ans = 1.7321
atan	$\tan^{-1}(x)$. Returns the output in radians	≫atan(1) ans = 0.7854
atand	$\tan^{-1}(x)$. Returns the output in degrees	≫atand(1) ans = 45
sec	$\sec(x)$. Takes the input in radians	≫sec(1) ans = 1.8508
secd	$\sec(x)$. Takes the input in degrees	≫secd(10) ans = 1.0154
asec	$\sec^{-1}(x)$. Returns the output in radians	≫asec(30) ans = 1.5375
asecd	$\sec^{-1}(x)$. Returns the output in degrees	≫asecd(30) ans = 88.0898
csc	$\csc(x)$. Takes the input in radians	≫csc(1) ans = 1.1884
cscd	$\csc(x)$. Takes the input in degrees	≫cscd(1) ans = 57.2987
acsc	$\csc^{-1}(x)$. Returns the output in radians	≫acsc(1) ans = 1.5708
acscd	$\csc^{-1}(x)$. Returns the output in degrees	≫acscd(1) ans = 90
cot	$\cot(x)$. Takes the input in radians	≫cot(1) ans = 0.6421
cotd	$\cot(x)$. Takes the input in degrees	≫cotd(30) ans = 1.7321
acot	$\cot^{-1}(x)$. Returns the output in radians	≫acot(1) ans = 0.7854
acotd	$\cot^{-1}(x)$. Returns the output in degrees	≫acotd(1) ans = 45

For instance, assume that we want to calculate for $\sin(x)^4 - 3\cos(x)$ for $x = 30°$. The commands shown in Figs. 1.44 or 1.45 do this.

Fig. 1.44 Calculation of $\sin(x)^4 - 3\cos(x)$ for $x = 30°$

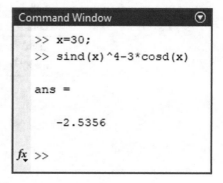

```
Command Window
   >> x=30*pi/180;
   >> sin(x)^4-3*cos(x)

   ans =

      -2.5356

fx >> |
```

Fig. 1.45 Calculation of $\sin(x)^4 - 3\cos(x)$ for $x = 30°$

```
Command Window
   >> x=30;
   >> sind(x)^4-3*cosd(x)

   ans =

      -2.5356

fx >>
```

1.7   Hyperbolic Functions

Hyperbolic functions of MATLAB are listed in Table 1.3. x can be a complex number as well.

Table 1.3 Hyperbolic functions of MATLAB

Function	MATLAB representation
$\sinh(x)$	`sinh(x)`
$\cosh(x)$	`cosh(x)`
$\tanh(x)$	`tanh(x)`
$\coth(x)$	`coth(x)`
$\text{sech}(x)$	`sech(x)`
$\text{csch}^{-1}(x)$	`csch(x)`
$\sinh^{-1}(x)$	`asinh(x)`
$\cosh^{-1}(x)$	`acosh(x)`
$\tanh^{-1}(x)$	`atanh(x)`
$\coth^{-1}(x)$	`acoth(x)`
$\text{sech}^{-1}(x)$	`asech(x)`
$\text{csch}^{-1}(x)$	`acsch(x)`

1.8   Logarithmic and Exponential Function

Logarithmic and exponential functions of MATLAB are listed in Table 1.4. x can be a complex number as well.

Table 1.4 Logarithmic and exponential functions of MATLAB

MATLAB function	Description	Example
exp	Calculates e^x	>>exp(2) ans = 7.3891
log	Calculates the natural logarithm (with base of $e = 2.7182$)	>>log(2.7182) ans = 1.0000
log10	Calculates the common logarithm (with base of 10)	>>log10(100) ans = 2
log2	Calculates the binary logarithm with (with base of 2)	>>log2(4) ans = 2
sqrt	Calculates the square root	>>sqrt(16) ans = 4
power	Calculates the power function. Same as elementwise power operator (Fig. 1.46)	>>power(2,3) ans = 8

Fig. 1.46 Result of power command

1.9   Rounding Functions

Rounding functions of MATLAB are shown in Table 1.5.

Table 1.5 Rounding function of MATLAB

MATLAB function	Expression	Example
fix	Rounds toward zero	fix([-4.6 4.6]) ans = [-4 4]
floor	Round to negative infinity	≫floor(8.4687) ans = 8
ceil	Round to positive infinity	≫ceil(4.4) ans = 5
round	Round to nearest decimal or integer	≫round(4.55) ans = 5

1.10   Colon Operator

The colon operator can be used for creating regularly spaced vectors, index into arrays, and define the bounds of a for loop. For instance, assume that you want to make a vector t from 0 to 1 with 0.1 s steps. The command shown in Fig. 1.47 makes the vecotor.

Fig. 1.47 Defining the vector
t with colon operator

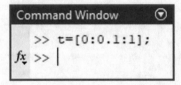

Values of vector t is shown in Fig. 1.48.

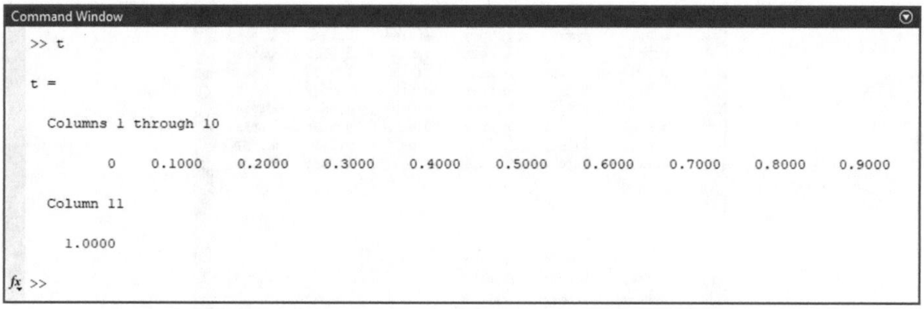

Fig. 1.48 Values of vector t defined in Fig. 1.47

The default step of colon operator equals to one. So, x = [3:10] and x = [3:1:10] commands are the same, both of these commands define x = [3 4 5 6 7 8 9 10]. The command shown in Fig. 1.49 defines a random matrix with 8 rows and 6 columns. The A(5:8,:) command in Fig. 1.49, selects the elements in the 5th, 6th, 7th and 8th columns.

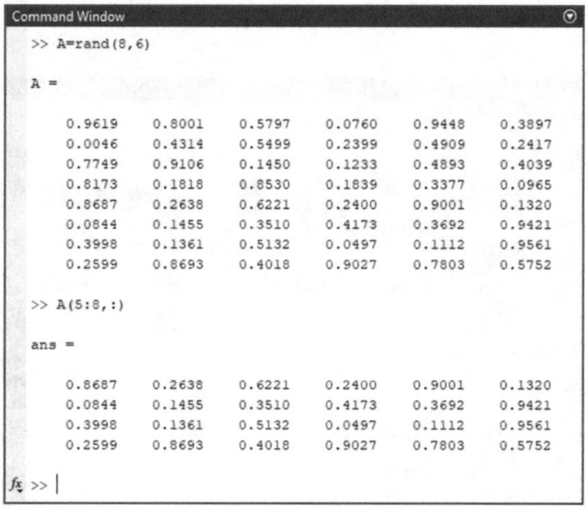

Fig. 1.49 Selection of 5th, 6th, 7th and 8th row of matrix A

The command A(:,4:6) in Fig. 1.50 selects all the elements in the 4th, 5th, 6th columns.

```
Command Window
>> A=rand(8,6)

A =

    0.9619    0.8001    0.5797    0.0760    0.9448    0.3897
    0.0046    0.4314    0.5499    0.2399    0.4909    0.2417
    0.7749    0.9106    0.1450    0.1233    0.4893    0.4039
    0.8173    0.1818    0.8530    0.1839    0.3377    0.0965
    0.8687    0.2638    0.6221    0.2400    0.9001    0.1320
    0.0844    0.1455    0.3510    0.4173    0.3692    0.9421
    0.3998    0.1361    0.5132    0.0497    0.1112    0.9561
    0.2599    0.8693    0.4018    0.9027    0.7803    0.5752

>> A(:,4:6)

ans =

    0.0760    0.9448    0.3897
    0.2399    0.4909    0.2417
    0.1233    0.4893    0.4039
    0.1839    0.3377    0.0965
    0.2400    0.9001    0.1320
    0.4173    0.3692    0.9421
    0.0497    0.1112    0.9561
    0.9027    0.7803    0.5752

fx >>
```

Fig. 1.50 Selection of 4th, 5th and 6th column of matrix A

The command A(2:5,3:4) in Fig. 1.51 selects the elements A_{ij} where $2 \leq i \leq 5$ and $3 \leq j \leq 4$.

```
Command Window
>> A=rand(8,6)

A =

    0.9619    0.8001    0.5797    0.0760    0.9448    0.3897
    0.0046    0.4314    0.5499    0.2399    0.4909    0.2417
    0.7749    0.9106    0.1450    0.1233    0.4893    0.4039
    0.8173    0.1818    0.8530    0.1839    0.3377    0.0965
    0.8687    0.2638    0.6221    0.2400    0.9001    0.1320
    0.0844    0.1455    0.3510    0.4173    0.3692    0.9421
    0.3998    0.1361    0.5132    0.0497    0.1112    0.9561
    0.2599    0.8693    0.4018    0.9027    0.7803    0.5752

>> A(2:5,3:4)

ans =

    0.5499    0.2399
    0.1450    0.1233
    0.8530    0.1839
    0.6221    0.2400

fx >>
```

Fig. 1.51 Selection of A_{ij} where $2 \leq i \leq 5$ and $3 \leq j \leq 4$

The colon operator is used in for loops as well. For instance, the code in Fig. 1.52 counts the number of elements which are bigger or equal than 0.5.

```
Command Window
>> A=rand(8,6)

A =

    0.9619    0.8001    0.5797    0.0760    0.9448    0.3897
    0.0046    0.4314    0.5499    0.2399    0.4909    0.2417
    0.7749    0.9106    0.1450    0.1233    0.4893    0.4039
    0.8173    0.1818    0.8530    0.1839    0.3377    0.0965
    0.8687    0.2638    0.6221    0.2400    0.9001    0.1320
    0.0844    0.1455    0.3510    0.4173    0.3692    0.9421
    0.3998    0.1361    0.5132    0.0497    0.1112    0.9561
    0.2599    0.8693    0.4018    0.9027    0.7803    0.5752

>> [m,n]=size(A);
>> s=0;
>> for i=1:m
for j=1:n
if (A(i,j)>=0.5)
s=s+1;
end
end
end
>> s

s =

    19

fx >>
```

Fig. 1.52 Counting the number of elements which are equal or bigger than 0.5

1.11   Linspace and Logspace Commands

The linspace(X1, X2, N) generates N linearly spaced points between X1 and X2. MAT-LAB assumes N = 100 when it is not entered. For instance, the command in Fig. 1.53 generates a linear space from 1 to 10 which has 6 members. We expect the difference between two consecutive elements to be $\frac{10-1}{6-1} = 1.8$.

```
Command Window
>> X=linspace(1,10,6)

X =

    1.0000    2.8000    4.6000    6.4000    8.2000   10.0000

fx >>
```

Fig. 1.53 Defining the variable X with linspace command

The difference between two consecutive elements can be calculated using the diff command (diff command can be used to calculate the derivative of a function as well). According to Fig. 1.54, the difference between two consecutive elements is 1.8.

```
Command Window                                                        ⊙
  >> X=linspace(1,10,6)

  X =

       1.0000     2.8000     4.6000     6.4000     8.2000    10.0000

  >> diff(X)

  ans =

       1.8000     1.8000     1.8000     1.8000     1.8000

fx >> |
```

Fig. 1.54 Difference between the consecutive elements of vector X

Logspace(X1, X2, N) generates a row vector of N logarithmically equally spaced points between 10^{X1} and 10^{X2}. MATLAB assumes N = 100 when it is not entered. For instance, the command in Fig. 1.55, makes a logarithmically equally spaced from 100 to 1000 with 5 members.

```
Command Window                                                        ⊙
  >> Y=logspace(2,3,5)

  Y =

       1.0e+03 *

         0.1000     0.1778     0.3162     0.5623     1.0000

fx >> |
```

Fig. 1.55 Defining the variable Y with logspace command

The ratio between any two consecutive elements are constant (Fig. 1.56). This ratio can be calculated from the $10^2 \times x^{5-1} = 10^3$ equation. Figure 1.57 solves this equation. According to Fig. 1.57, the only positive answer of this equation is 1.7783.

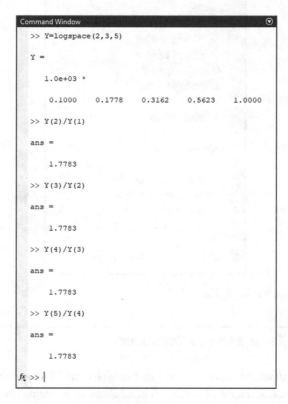

Fig. 1.56   Ratio between the consecutive elements of vector Y

```
Command Window                    ⊙
  >> solve(100*x^4==1000)

  ans =

        10^(1/4)
       -10^(1/4)
     -10^(1/4)*1i
      10^(1/4)*1i

  >> eval(ans)

  ans =

        1.7783 + 0.0000i
       -1.7783 + 0.0000i
        0.0000 - 1.7783i
        0.0000 + 1.7783i

fx >> |
```

Fig. 1.57 Solution of $10^2 \times x^{5-1} = 10^3$

1.12 Ones, Zeros and Eye Commands

The ones(m, n) command makes a m × n matrix which all of its elements are one. The zeros (m, n) makes a m × n matrix which all of its elements are zero. The eye(n) command makes a n × n identity matrix. The eye(m, n) command makes a m × n matrix with 1's on the diagonal and 0 elsewhere (Fig. 1.58).

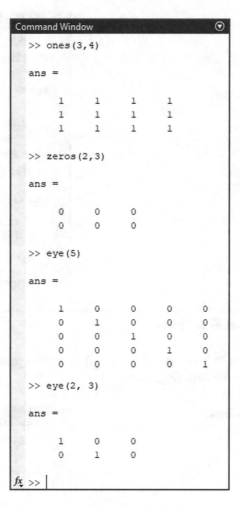

Fig. 1.58 Example for ones, zeros and eye commands

1.13   Format Command

The format command permits you to set the output display format for command window (Fig. 1.59).

Fig. 1.59 The format
command

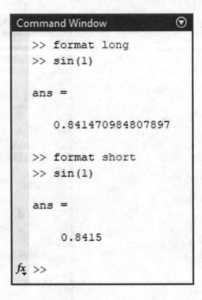

1.14   Polynomial Functions

Assume that you want to solve the $x^2 - 4x + 3 = 0$. The roots command takes the coefficients and return the roots (Fig. 1.60).

Fig. 1.60 The roots command
calculates the roots of a
polynomial

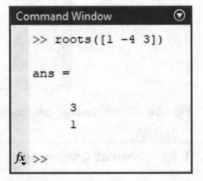

Assume that you want to find the polynomial which its roots are $-1 + 2i, -1 - 2i$ and 6. The poly function gives the coefficients of the polynomial which has that roots. According to Fig. 1.61, the polynomial is $p(x) = x^3 - 4x^2 - 7x - 30$.

Fig. 1.61 Poly command
takes the roots of a polynomial
function and gives its
coefficients

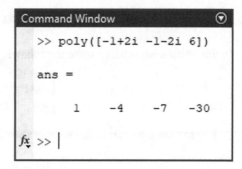

Assume that we want to calculate the value of $p(x) = x^3 - 4x^2 - 7x - 30$ at $x = 5$. The polyval function can be used for this purpose (Fig. 1.62).

Fig. 1.62 Polyval command
calculates the value of a
polynomial function at a point

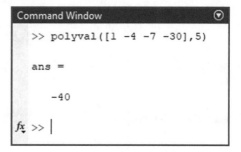

Assume that you want to calculate the product of $p_1(x) = x + 3$ and $p_2(x) = x^2 + 5x + 6$. The commands shown in Fig. 1.63 calculates the product of these two polynomials. According to Fig. 1.63, the product is $p(x) = x^3 + 8x^2 + 21x + 18$.

Fig. 1.63 Calculation of
product of two polynomials
with the conv command

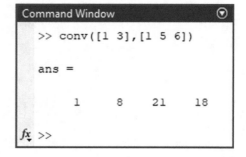

1.15   Solution of Nonlinear Systems

In this section we want to solve the following nonlinear system

$$\begin{cases} 2x_1 - x_2 = e^{-x_1} \\ -x_1 + 2x_2 = e^{-x_2} \end{cases} \tag{1.1}$$

Rewrite the equation in the form F(x) = 0:

$$\begin{cases} 2x_1 - x_2 - e^{-x_1} = 0 \\ -x_1 + 2x_2 - e^{-x_2} = 0 \end{cases} \tag{1.2}$$

Use the edit command in order to run the MATLAB editor (Fig. 1.64) (Fig. 1.65).

Fig. 1.64 The edit command

Fig. 1.65 Editor is opened

Enter the code shown in Fig. 1.66 and press the Ctrl+S to save it. This code solves the nonlinear system with initial guess of [− 5, − 5]. This code asks the MATLAB to show the iterations.

```
Editor - D:\00Numerical Analysis with MATLAB\Programs\untitled.m
untitled.m

F = @(x) [2*x(1) - x(2) - exp(-x(1));
          -x(1) + 2*x(2) - exp(-x(2))];
x0=[-5,-5];
options=optimoptions('fsolve','Display','iter');
[x,fval]=fsolve(F,x0,options);
```

Fig. 1.66 Code to solve the given system

Run the code by pressing the F5 key of your keyboard. Result is shown in Fig. 1.67.

```
Command Window
>> untitled

                                         Norm of      First-order   Trust-region
 Iteration  Func-count     f(x)          step         optimality    radius
     0          3         47071.2                      2.29e+04               1
     1          6         12003.4              1       5.75e+03               1
     2          9         3147.02              1       1.47e+03               1
     3         12         854.452              1            388               1
     4         15         239.527              1            107               1
     5         18         67.0412              1           30.8               1
     6         21         16.7042              1           9.05               1
     7         24         2.42788              1           2.26               1
     8         27        0.032658       0.759511          0.206             2.5
     9         30     7.03149e-06       0.111927        0.00294             2.5
    10         33     3.29525e-13     0.00169132       6.36e-07             2.5

Equation solved.

fsolve completed because the vector of function values is near zero
as measured by the value of the function tolerance, and
the problem appears regular as measured by the gradient.

<stopping criteria details>
>>
```

Fig. 1.67 Output of the code in Fig. 1.66

Use the commands shown in Fig. 1.68 in order to see the solution.

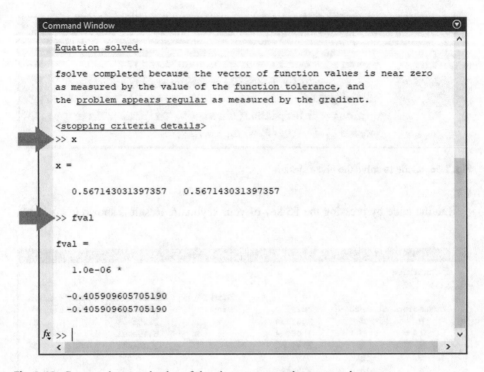

Fig. 1.68 Computed root and value of the given system at the computed root

Let's study another example. We want to solve the following system.

$$\begin{cases} e^{-e^{-(x_1+x_2)}} = x_2(1 + x_1^2) \\ x_1 \cos(x_2) + x_2 \sin(x_1) = \frac{1}{2} \end{cases} \tag{1.3}$$

Let's convert the equations to F(x) = 0 form.

$$\begin{cases} e^{-e^{-(x_1+x_2)}} - x_2(1 + x_1^2) = 0 \\ x_1 \cos(x_2) + x_2 \sin(x_1) - \frac{1}{2} = 0 \end{cases} \tag{1.4}$$

Enter the code shown in Fig. 1.69 and press the Ctrl+S to save it. This code solves the nonlinear system with initial guess of [0, 0]. This code doesn't ask the MATLAB to show the iterations.

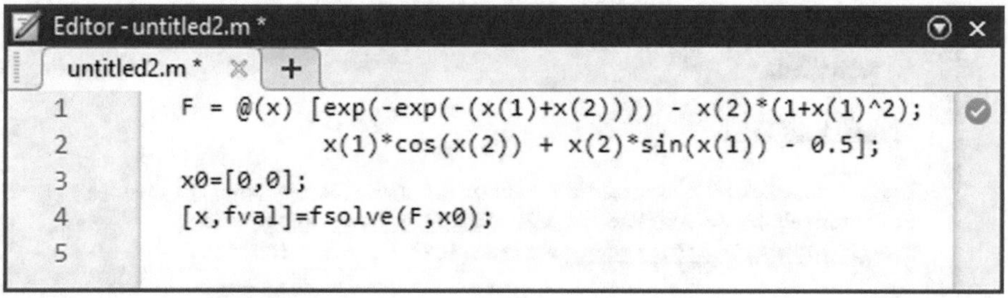

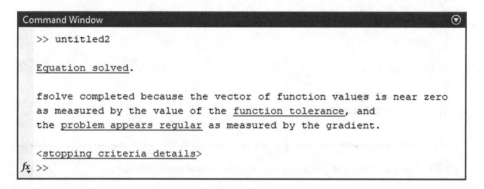

Fig. 1.69 Code to solve the given system

Run the code by pressing the F5 key of your keyboard. Result is shown in Fig. 1.70.

Fig. 1.70 Output of the code in Fig. 1.69

Use the commands shown in Fig. 1.71 in order to see the solution.

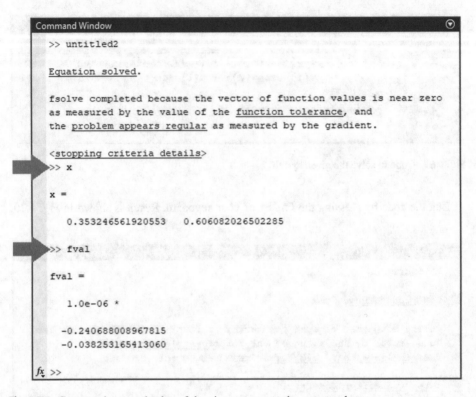

Fig. 1.71 Computed root and value of the given system at the computed root

1.16   Eigen Values and Eigen Vectors

You can use the eig function to calculate the eigen values and eigen vectors of a matrix.

For instance, for $A = \begin{bmatrix} 5 & 7 & -5 \\ 0 & 4 & -1 \\ 2 & 8 & -3 \end{bmatrix}$,

$$\begin{bmatrix} 5 & 7 & -5 \\ 0 & 4 & -1 \\ 2 & 8 & -3 \end{bmatrix} \times \begin{bmatrix} -0.5345 \\ -0.2673 \\ -0.8018 \end{bmatrix} = 1 \times \begin{bmatrix} -0.5345 \\ -0.2673 \\ -0.8018 \end{bmatrix} = \begin{bmatrix} -0.5345 \\ -0.2673 \\ -0.8018 \end{bmatrix}$$

$$\begin{bmatrix} 5 & 7 & -5 \\ 0 & 4 & -1 \\ 2 & 8 & -3 \end{bmatrix} \times \begin{bmatrix} -0.4082 \\ -0.4082 \\ -0.8165 \end{bmatrix} = 2 \times \begin{bmatrix} -0.4082 \\ -0.4082 \\ -0.8165 \end{bmatrix} = \begin{bmatrix} -0.8164 \\ -0.8164 \\ -0.8164 \end{bmatrix}$$

$$\begin{bmatrix} 5 & 7 & -5 \\ 0 & 4 & -1 \\ 2 & 8 & -3 \end{bmatrix} \times \begin{bmatrix} -0.5774 \\ 0.5774 \\ 0.5774 \end{bmatrix} = 3 \times \begin{bmatrix} -0.5774 \\ 0.5774 \\ 0.5774 \end{bmatrix} = \begin{bmatrix} -1.7322 \\ 1.7322 \\ 1.7322 \end{bmatrix} \tag{1.5}$$

Enter the commands shown in Fig. 1.72. As you see, the values of eigen vectors will be saved in the matrix V and eigen values will be saved in the matrix D.

Fig. 1.72 Eigen values are calculated with the aid of eig command

```
Command Window

>> A=[5 7 -5;0 4 -1;2 8 -3];
>> [V,D]=eig(A)

V =

    -0.5345   -0.5774   -0.4082
    -0.2673    0.5774   -0.4082
    -0.8018    0.5774   -0.8165


D =

    1.0000         0         0
         0    3.0000         0
         0         0    2.0000

fx >>
```

1.17 Reduced Echelon Form of a Matrix

You can obtain the reduced echelon form of matrix with the aid of rref command. For instance, consider the following system of equations

$$\begin{cases} 5x + 7y - 5z = 0 \\ 0x + 4y - z = 4 \\ 2x + 8y - 3z = 2 \end{cases} \tag{1.6}$$

The reduced echelon form of $\begin{bmatrix} 5 & 7 & -5 & 0 \\ 0 & 4 & -1 & 4 \\ 2 & 8 & -3 & 2 \end{bmatrix}$ can be obtained with the aid of commands shown in Fig. 1.73.

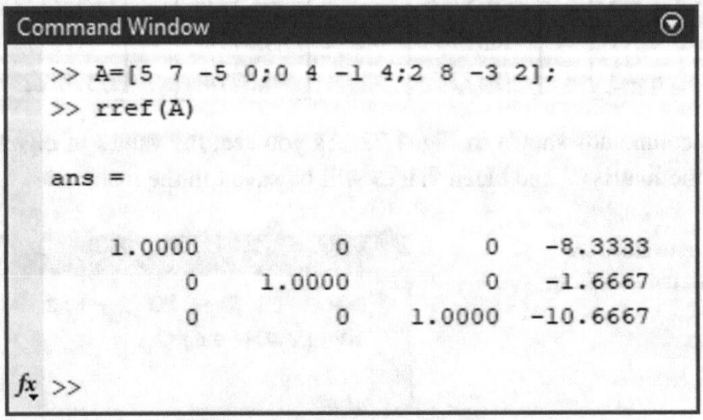

```
Command Window                                              ⊙

  >> A=[5 7 -5 0;0 4 -1 4;2 8 -3 2];
  >> rref(A)

  ans =

      1.0000         0         0   -8.3333
           0    1.0000         0   -1.6667
           0         0    1.0000  -10.6667

fx >>
```

Fig. 1.73 rref can be used to compute the reduced echelon form of a matrix

Let's solve the given system and obtain the solution. This can be done with the aid of commands shown in Fig. 1.74. Result shown in Fig. 1.74 is the same as the fourth column of the reduced echelon form matrix.

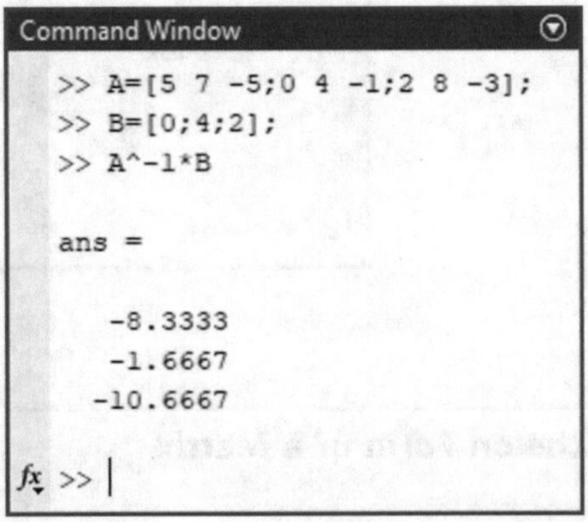

```
Command Window                                              ⊙

  >> A=[5 7 -5;0 4 -1;2 8 -3];
  >> B=[0;4;2];
  >> A^-1*B

  ans =

      -8.3333
      -1.6667
     -10.6667

fx >> |
```

Fig. 1.74 Solution of the given linear system

1.18   Norm of Vectors and Matrices

You can compute the p-norm of a vector with the aid of norm command (Fig. 1.75). The p-norm of a vector $\begin{bmatrix} x_1 \ x_2 \ x_3 \ \dots \ x_n \end{bmatrix}$ is defined as $\sqrt[p]{\sum_{i=1}^{n} |x_i|^p}$. ∞-norm returns the max$\{|x_1|, |x_2|, |x_3|, \dots, |x_n|\}$.

Fig. 1.75 norm command

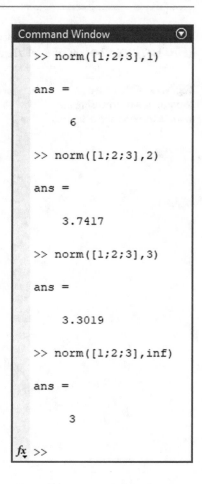

For instance, let's compute the 3-norm of the vector shown in Fig. 1.75 and compare it with the result shown in Fig. 1.76. According to Fig. 1.76, the result obtained in Fig. 1.75 is correct.

Fig. 1.76 Computation of 3-norm for the [1–3] vector

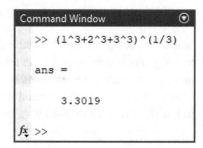

You can compute the 2-norm of a matrix with the aid of the norm function as well (Fig. 1.77). The 2-norm of a matrix A is defined as $\|A_2\| = \sqrt{\text{max eigen value of } A^T A}$.

Fig. 1.77 Calculation of 2-norm of a matrix using the norm command and definition

```
Command Window

>> A=[5 7 -5;0 4 -1;2 8 -3]

A =

        5       7      -5
        0       4      -1
        2       8      -3

>> norm(A)

ans =

    13.5469

>> sqrt(max(eig(A'*A)))

ans =

    13.5469

fx >>
```

1.19   Condition Number of a Matrix

The condition number of a matrix A is defined as

$$k(A) = \|A\|.\|A^{-1}\|$$

where $\|A\|$ is the norm of the matrix defined above. Matrices with condition number close to unity are said to be well–conditioned matrices, and those with very large condition number are said to be ill–conditioned matrices. The condition number of a matrix is computed with the cond command. For instance, condition number of the matrix A shown in Fig. 1.77 is equal to 94.0730 (Fig. 1.78).

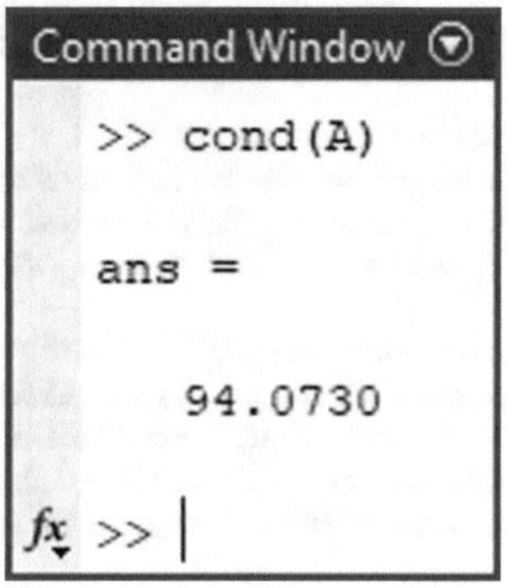

Fig. 1.78 Computation of condition number of a matrix

1.20 Tic and Toc Commands

The tic and toc commands are used to measure the elapsed time. For instance, according to Fig. 1.79, the required time to run the roots command is 0.038868 s. The required time to run a command depends on the hardware (i.e. RAM, CPU, …) of your computer. So, the number that appear on your computer may be different.

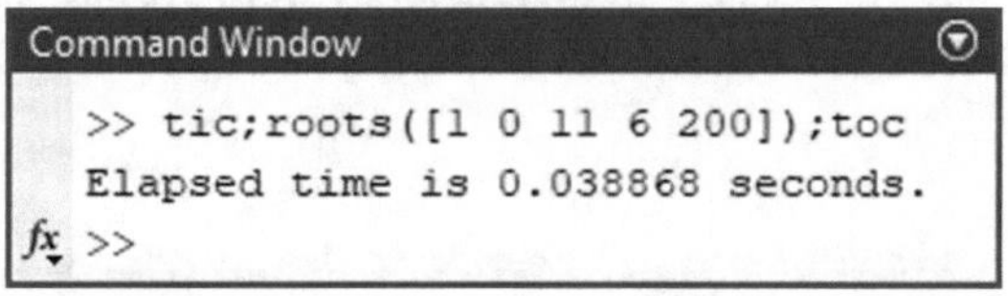

Fig. 1.79 Obtaining the elapsed time

1.21 Getting Help in MATLAB

You can use the help command to get more information about the commands studied in this chapter. For instance, assume that you need more information about the roots command. The help roots command brings the document of the command for you (Fig. 1.80).

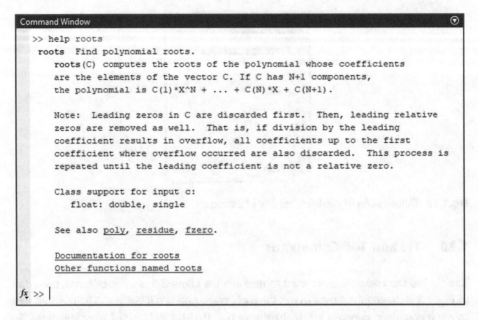

```
Command Window                                                               ⊙
  >> help roots
  roots  Find polynomial roots.
    roots(C) computes the roots of the polynomial whose coefficients
    are the elements of the vector C. If C has N+1 components,
    the polynomial is C(1)*X^N + ... + C(N)*X + C(N+1).

    Note:  Leading zeros in C are discarded first.  Then, leading relative
    zeros are removed as well.  That is, if division by the leading
    coefficient results in overflow, all coefficients up to the first
    coefficient where overflow occurred are also discarded.  This process is
    repeated until the leading coefficient is not a relative zero.

    Class support for input c:
      float: double, single

    See also poly, residue, fzero.

    Documentation for roots
    Other functions named roots

fx >> |
```

Fig. 1.80 Output of help roots command

There are other ways to get help in MATLAB as well. For instance, if you press the F1 key of the keyboard, the window shown in Fig. 1.81 appears. Click the open help browser.

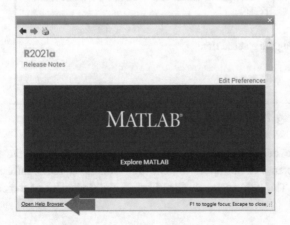

Fig. 1.81 MATLAB help window

After clicking the open help browser (Fig. 1.81), the window shown in Fig. 1.82 appears. Enter the search term into the search documentation box. As you type the search term, a list of related topics appears below the search box. You click on the appeared topics to see their explanation.

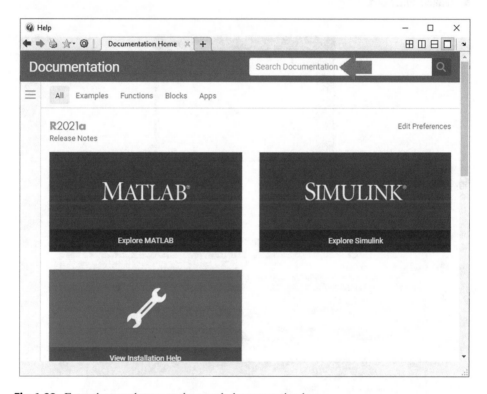

Fig. 1.82 Enter the search term to the search documentation box

Let's study another way to get help. On the left side, inside the command window, there is something visible in small fonts, and that is fx (Fig. 1.83). If you click on it (or press Shift+F1), a drop-down search bar gets opened (Fig. 1.84). The appeared list is the name of the products you have installed. Click on any of the products to get a list of all the related functions.

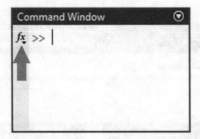

Fig. 1.83 The fx button

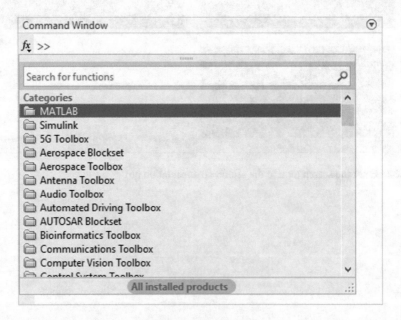

Fig. 1.84 The window which appears after clicking the fx button

Exercises

1. Do the following calculations with MATLAB.
   (a) $\sin(1 + 2i)$ (b) e^{1+2i} (c) $\text{Log}_{10}(1 + 2i)$ (d) $Ln(1 + 2i)$ (e) $\frac{1+2i}{3+4i}$ (f) $\sin(60°)$ (g) $\cos(\frac{\pi}{3})$ (h) $\sqrt{8^2 + 2^2}$ (i) $\sqrt[4]{8^2 + 2^2}$.

2. Solve the $x^4 + 2x^3 + 4x + 1 = 0$ with MATLAB.

3. Solve the following linear system of equations with MATLAB.

$$\begin{cases} 5x + 17y - 2z = 0 \\ 1x + 3y - z = 4 \\ 2x + 8y - 4z = 20 \end{cases}$$

4. Solve the following nonlinear system of equations with MATLAB.

$$\begin{cases} 3x_1 - x_2 = e^{-x_1} \\ -2x_1 + 2x_2 = e^{-x_2} \end{cases}$$

5. Calculate the determinant and inverse of $\begin{bmatrix} 8 & 7 & -1 \\ 8 & 4 & -1 \\ 0 & 8 & -6 \end{bmatrix}$ with MATLAB.

6. Write a MATLAB code to calculate the product of odd numbers from 1 to 10. **Hint:** Use the prod command.

References for Further Study

1. Chapman, S.: MATLAB Programming for Engineers, 6th edition, Cengage, 2019.
2. Hahn, B., Valentine, D.: Essential MATLAB for Engineers and Scientists, 7th edition, Academic Press, 2019.
3. Moore, H.: MATLAB for Engineers, 5th edition, Pearson, 2017.

Symbolic Calculations in MATLAB®

2

2.1 Introduction

Symbolic Math Toolbox™ provides functions for solving, plotting, and manipulating symbolic math equations. In this chapter you will learn how to do symbolic calculations in MATLAB. You will learn how to define a symbolic variable, how to calculate limit, derivative and integral, how to solve a differential equation, how to draw the graph of a function, how to obtain the Laplace and Fourier transforms of a function, and how to calculate the Taylor series of a function.

2.2 Calculation of Limit, Derivative and Integral

Table 2.1 shows how to calculate the limit, derivative and integral using MATLAB. The syms command creates a symbolic variable for the calculations.

© The Author(s), under exclusive license to Springer Nature Switzerland AG 2023
F. Asadi, *Applied Numerical Analysis with MATLAB®/Simulink®*,
Synthesis Lectures on Engineering, Science, and Technology,
https://doi.org/10.1007/978-3-031-19366-8_2

Table 2.1 Example for lim, diff and int functions of MATLAB

Mathematical expression	MATLAB representation
$\lim\limits_{x \to 0} \frac{\sin(x)}{x} = ?$	>>syms x >>limit(sin(x)/x,0)
$f(x) = \frac{x^3 - x}{x^2 + 8}$ $\frac{df}{dx} = ?$	>>syms x >>diff((x^3-x)/(x^2 + 8))
$f(x, y) = x^3 + 3y^2 + 4xy$ $\frac{\partial f}{\partial y} = ?$	>>syms x y >>f = x^3 + 3*y^2 + 4*x*y >>diff(f,y)
$\int \frac{x^3 - x}{x^2 + 8} dx = ?$	>>syms x >>int((x^3-x)/(x^2 + 8))
$\int\limits_0^1 \frac{x^3 - x}{x^2 + 8} dx = ?$	>>syms x >>int((x^3-x)/(x^2 + 8),x,0,1)

The command shown in Fig. 2.1 calculates the second derivative of function f with respect to variable x.

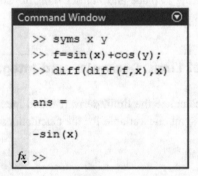

Fig. 2.1 Calculation of $\frac{d^2(\sin(x) + \cos(y))}{dx^2}$

Sometimes it is difficult to read the MATLAB result since MATLAB shows the result in only one line. In these cases, you can use the pretty function which shows the result in a more readable manner (Fig. 2.2).

```
Command Window                    ⊙
>> syms x
>> int((x^3-x)/(x^2+8))

ans =

x^2/2 - (9*log(x^2 + 8))/2

>> pretty(ans)
  2              2
 x      9 log(x  + 8)
 --  -  -------------
  2           2

fx >> |
```

Fig. 2.2 Calculation of integral with int command

The simplify command, can be used to simplify the result of symbolic calculation (Fig. 2.3).

```
Command Window                    ⊙
>> syms x
>> simplify(1/(x+1)+1/(x-1))

ans =

(2*x)/(x^2 - 1)

>> pretty(ans)
   2 x
  ------
   2
  x  - 1

fx >> |
```

Fig. 2.3 Simplify and pretty commands

2.3 Solving the Ordinary Differential Equations

Assume that you want to solve $\frac{dy(t)}{dt} + 4y(t) = e^{-t}$, $y(0) = 1$. The commands shown in Fig. 2.4 solve this differential equation. Note that the first line defines the symbolic function y as a function of symbolic variable t.

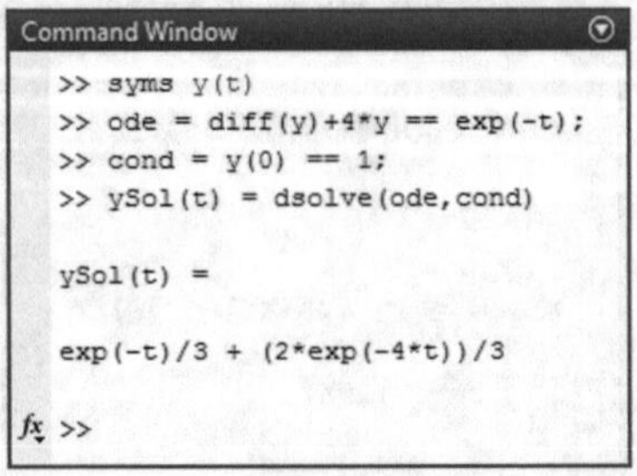

```
Command Window                                    ⊙
>> syms y(t)
>> ode = diff(y)+4*y == exp(-t);
>> cond = y(0) == 1;
>> ySol(t) = dsolve(ode,cond)

ySol(t) =

exp(-t)/3 + (2*exp(-4*t))/3

fx >>
```

Fig. 2.4 Solution of $\frac{dy(t)}{dt} + 4y(t) = e^{-t}$, $y(0) = 1$

Let's study another example. Assume that we want to solve $2\frac{d^2y(t)}{dt^2} + \frac{dy(t)}{dt} + 11y(t) = e^{-t}$, $y(0) = 1$, $y'(0) = 0$. The commands shown in Fig. 2.5 solves this problem.

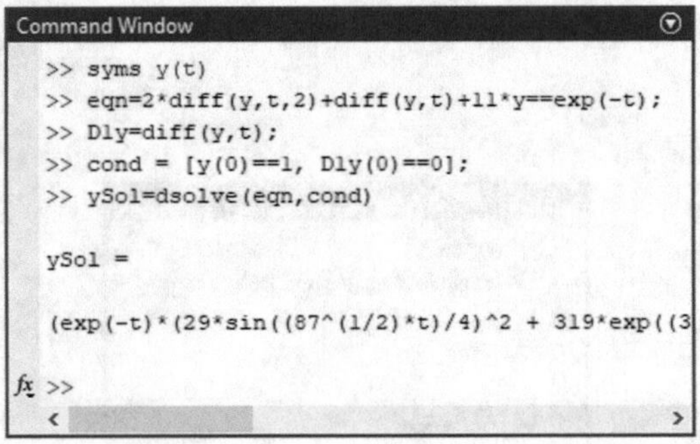

```
Command Window                                    ⊙
>> syms y(t)
>> eqn=2*diff(y,t,2)+diff(y,t)+11*y==exp(-t);
>> D1y=diff(y,t);
>> cond = [y(0)==1, D1y(0)==0];
>> ySol=dsolve(eqn,cond)

ySol =

(exp(-t)*(29*sin((87^(1/2)*t)/4)^2 + 319*exp((3

fx >>
<                                              >
```

Fig. 2.5 Solution of $2\frac{d^2y(t)}{dt^2} + \frac{dy(t)}{dt} + 11y(t) = e^{-t}$, $y(0) = 1$, $y'(0) = 0$

Assume that we want to see the graph of ySol for [0 s, 6 s] time interval. You can use fplot command (Fig. 2.6) to see the ySol graph (Fig. 2.7).

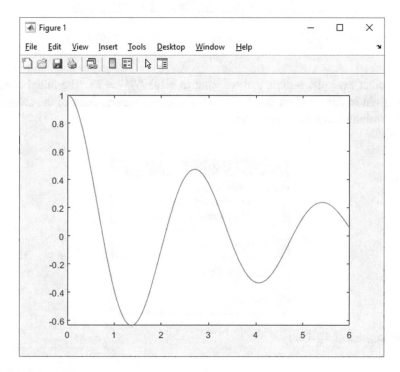

Fig. 2.6 Drawing the obtained solution with fplot command

Fig. 2.7 Result of fplot command

Assume that we want to solve $\frac{d^3y(x)}{dx^3} = \cos(2x) - y$, $y(0) = 1$, $y'(0) = 0$, $y''(0) = -1$. The commands shown in Fig. 2.8 solve this problem. You can use the simplify function to simplify the calculated result.

```
Command Window                                                          ⊙
  >> syms y(x)
  >> eqn=diff(y,x,2)==cos(2*x)-y;
  >> D1y=diff(y,x);
  >> D2y=diff(y,x,2);
  >> cond = [y(0)==1, D1y(0)==0, D2y(0)==0];
  >> ySol=dsolve(eqn,cond)

  ySol =

  (5*cos(x))/3 + sin(x)*(sin(3*x)/6 + sin(x)/2) -

  >> simplify(ySol)

  ans =

  1 - (8*sin(x/2)^4)/3

fx >> |
  <                                                                   >
```

Fig. 2.8 Solution of $\frac{d^3y(x)}{dx^3} = \cos(2x) - y$, $y(0) = 1$, $y'(0) = 0$, $y''(0) = -1$

As another example, assume that we want to solve $\frac{dy(t)}{dt} = ty$. The initial conditions are not given in this problem. So, MATLAB shows the general form of the solution. C1 shows an arbitrary constant (Fig. 2.9).

```
Command Window                              ⊙
  >> syms y(t)
  >> eqn=diff(y,t)==t*y;
  >> ySol=dsolve(eqn)

  ySol =

  C1*exp(t^2/2)

fx >>
```

Fig. 2.9 Solution of $\frac{dy(t)}{dt} = ty$

2.4 Partial Fraction Expansion and Laplace Transform

You can use the residue command to calculate the partial fraction expansions. For instance, assume that we want to write the partial fraction expansion of $\frac{s+1}{s^2+5s+6}$. The commands shown in Fig. 2.10 calculates the residues at the poles. So, according to Fig. 2.10, the partial fraction expansion of $\frac{s+1}{s^2+5s+6}$ equals to $\frac{2}{s+3} + \frac{-1}{s+2}$.

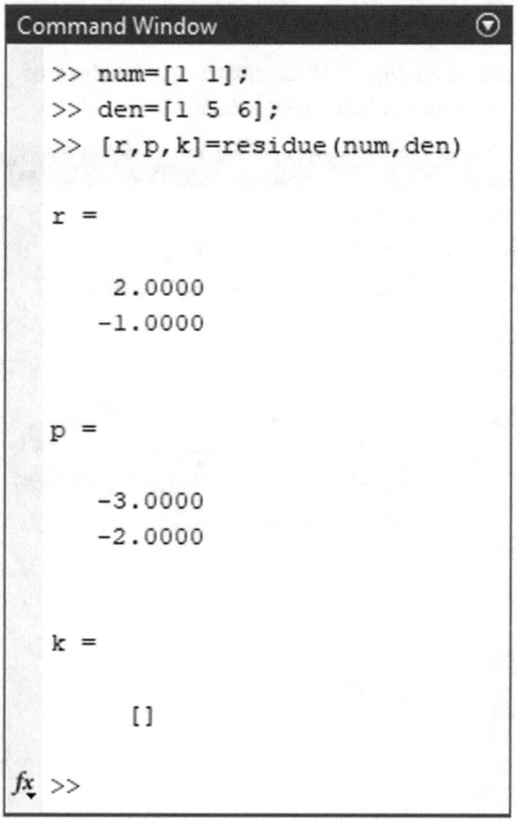

```
Command Window                           ⊙

>> num=[1 1];
>> den=[1 5 6];
>> [r,p,k]=residue(num,den)


r =

        2.0000
       -1.0000


p =

       -3.0000
       -2.0000


k =

        []

fx >>
```

Fig. 2.10 Result of residue command for $\frac{s+1}{s^2+5s+6}$

Let's study another example. Assume that we want to write the partial fraction expansion of $\frac{1}{s^3(s-0.5)}$. Using the residue theory, $\frac{1}{s^3(s-0.5)} = \frac{-2}{s^3} + \frac{-4}{s^2} + \frac{-8}{s} + \frac{8}{s-0.5}$. Figure 2.11 shows that the result is correct.

Fig. 2.11 Verification of
$\frac{-2}{s^3} + \frac{-4}{s^2} + \frac{-8}{s} + \frac{8}{s-0.5} =$
$\frac{1}{s^3(s-0.5)}$

```
Command Window

>> syms s
>> simplify(-2/s^3-4/s^2-8/s+8/(s-0.5))

ans =

2/(s^3*(2*s - 1))

fx >>
```

The commands shown in Fig. 2.12 calculates the residues of $\frac{1}{s^3(s-0.5)} = \frac{1}{s^4-0.5s^3}$. Obtained results are the same as hand calculations.

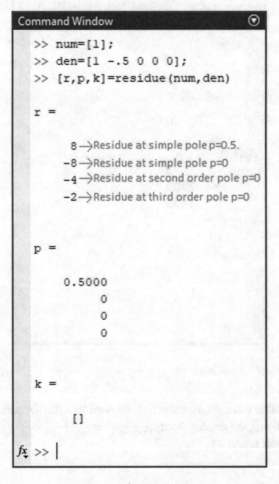

```
Command Window

>> num=[1];
>> den=[1 -.5 0 0 0];
>> [r,p,k]=residue(num,den)

r =

        8→Residue at simple pole p=0.5.
       -8→Residue at simple pole p=0
       -4→Residue at second order pole p=0
       -2→Residue at third order pole p=0


p =

    0.5000
         0
         0
         0


k =

    []

fx >> |
```

Fig. 2.12 Result of residue command for $\frac{1}{s^3(s-0.5)}$

The matrix k was a null matrix in both of the above examples. The matrix k is always null when the degree of the numerator is less than the denominator. When the degree of the numerator is equal or bigger than the denominator, it will not be null. For instance, assume $\frac{s^3+3s^2+7s+4}{s^2+2s+1}$. The commands shown in Fig. 2.13 calculates the partial fraction of this function. Note that matrix k is not null since the degree of numerator is bigger than denominator. According to Fig. 2.13, $\frac{s^3+3s^2+7s+4}{s^2+2s+1} = s + 1 + \frac{4}{(s+1)} + \frac{-1}{(s+1)^2}$.

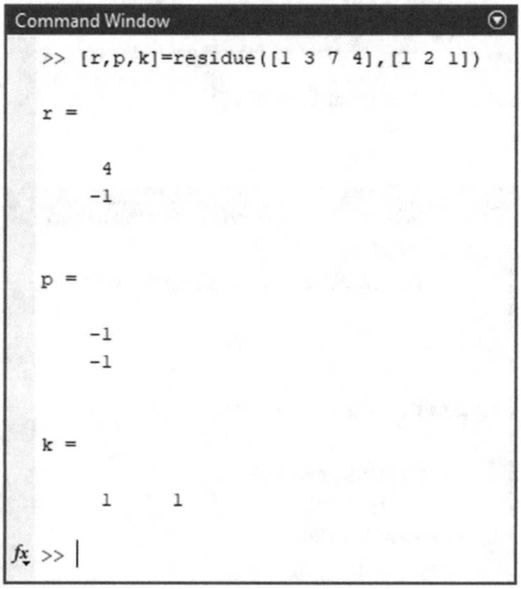

```
Command Window                                    ⊙

  >> [r,p,k]=residue([1 3 7 4],[1 2 1])

  r =

            4
           -1


  p =

           -1
           -1


  k =

           1    1

fx >> |
```

Fig. 2.13 Result of residue command for $\frac{s^3+3s^2+7s+4}{s^2+2s+1}$

You can calculate the Laplace transform of function quite easily. The commands shown in Figs. 2.14 and 2.15 calculates the Laplace transform of e^{-6t} and $e^{-6t}\sin(\omega t)$, respectively.

```
Command Window                              ⊙
    >> syms t
    >> laplace(exp(-6*t))

    ans =

    1/(s + 6)

fx >> |
```

Fig. 2.14 Calculation of the Laplace transform of e^{-6t}

```
Command Window                              ⊙
    >> syms w t
    >> laplace(exp(-6*t)*sin(w*t))

    ans =

    w/((s + 6)^2 + w^2)

    >> pretty(ans)
          w
    -------------
            2    2
    (s + 6)  + w

fx >> |
```

Fig. 2.15 Calculation of the Laplace transform of $e^{-6t}\sin(\omega t)$

You can use the ilaplace command to calculate the inverse Laplace transform. For instance, the commands shown in Fig. 2.16 calculates the inverse Laplace transform of $\frac{s+1}{s^2+2s+1}$. Defining the variable F is not mandatory, you can calculate the inverse Laplace transform directly as well (Fig. 2.17).

```
Command Window
    >> syms s
    >> F=(s+1)/(s^2+2*s+2);
    >> ilaplace(F)

    ans =

    exp(-t)*cos(t)

fx >>
```

Fig. 2.16 Calculation of inverse Laplace of $\frac{s+1}{s^2+2s+1}$

```
Command Window
    >> ilaplace((s+1)/(s^2+2*s+2))

    ans =

    exp(-t)*cos(t)

fx >>
```

Fig. 2.17 Calculation of inverse Laplace of $\frac{s+1}{s^2+2s+1}$

2.5 Fourier Transform

The Fourier transform of the expression $f = f(x)$ with respect to the variable x at the point w is

$$F(w) = c \int_{-\infty}^{+\infty} f(x)e^{iswx}dx$$

c and s are parameters of the Fourier transform. The Fourier function uses $c = 1$ and $s = -1$. Commands shown in Fig. 2.18 compute the Fourier transform of te^{-t^2} with $c = 1$ and $s = -1$.

```
Command Window                                    ▼
  >> syms t w
  >> fourier(t*exp(-t^2),t,w)

  ans =

  -(w*pi^(1/2)*exp(-w^2/4)*1i)/2

  >> pretty(ans)
                        /    2 \
                        |   w  |
     w sqrt(pi) exp|  - -- | 1i
                        \    4 /
    - ---------------------------
                   2
fx >>                                             ⌄
```

Fig. 2.18 Fourier transform of te^{-t^2} with $c = 1$ and $s = -1$

The commands shown in Fig. 2.19 calculates the invers Fourier of function obtained in Fig. 2.18.

```
Command Window                                    ▼
  >> syms t w
  >> F=fourier(t*exp(-t^2),t,w)

  F =

  -(w*pi^(1/2)*exp(-w^2/4)*1i)/2

  >> ifourier(F,w,t)

  ans =

  t*exp(-t^2)

fx >>
```

Fig. 2.19 Inverse Fourier transform of Fig. 2.18

Commands shown in Fig. 2.20 compute the Fourier transform of te^{-t^2} with $c = \frac{1}{\sqrt{2}}$ and $s = -1$.

```
Command Window                                              ⊙
   >> sympref('FourierParameters',[1/sqrt(2) -1]);
   >> fourier(t*exp(-t^2),t,w)

   ans =

   -(2^(1/2)*w*pi^(1/2)*exp(-w^2/4)*1i)/4

   >> pretty(ans)
                              /    2 \
                              |   w  |
         sqrt(2) w sqrt(pi) exp| - -- | 1i
                              \   4 /
      - ------------------------------------
                         4
fx >>
```

Fig. 2.20 Fourier transform of te^{-t^2} with $c = \frac{1}{\sqrt{2}}$ and $s = -1$

2.6 Taylor Series

The commands shown in Fig. 2.21 calculates the Taylor series of e^x at $x = 0$. The Taylor commands calculates the Taylor series up to 6 terms. If you need more or less terms, you can use the 'Order' option.

```
Command Window                                                    ⊙

   >> syms x
   >> T1=taylor(exp(x))

   T1 =

   x^5/120 + x^4/24 + x^3/6 + x^2/2 + x + 1

   >> T1=taylor(exp(x),'Order',7)

   T1 =

   x^6/720 + x^5/120 + x^4/24 + x^3/6 + x^2/2 + x + 1

fx >> |
```

Fig. 2.21 Taylor series of e^x at $x = 0$

The commands shown in Fig. 2.22 calculates the Taylor series of e^x at $x = 1$.

```
Command Window                                                    ⊙

   >> syms x
   >> T2=taylor(sin(x),x,1)

   T2 =

   sin(1) - (sin(1)*(x - 1)^2)/2 + (sin(1)*(x - 1)^4)/24 + cos(1)*(

   >> pretty(T2)
                              2                    4
                 sin(1) (x - 1)       sin(1) (x - 1)
   sin(1) - ---------------- + ---------------- + cos(1) (x - 1)
                     2                    24


                     3                    5
        cos(1) (x - 1)       cos(1) (x - 1)
      - ---------------- + ----------------
               6                    120

fx >> |

<                                                                  >
```

Fig. 2.22 Taylor series of $\sin(x)$ at $x = 1$

The code shown in Fig. 2.23 draws the graph of sin(x) and the first 5 terms of the Taylor series on the same axis. Output of this code is shown in Fig. 2.24.

```
Command Window                                              ⊙
  >> syms x
  >> y=sin(x);
  >> T5=taylor(sin(x),x,'Order',5);
  >> fplot([y T5]), xlim([-pi pi])
  >> grid on
  >> legend('5th order taylor approximation','sin(x)')
ƒx >>
```

Fig. 2.23 Comparison of sin(*x*) and the first 5 terms of the Taylor series

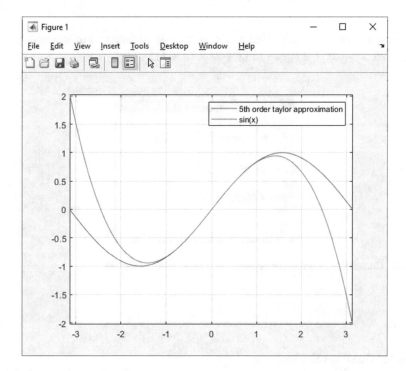

Fig. 2.24 Output of the code in Fig. 2.23

2.7 Expansion of an Algebraic Expression

You can expand an expression with the aid of expand command. An example is given in Fig. 2.25.

```
Command Window                          ⌄

    >> syms x
    >> expand((x+1)*(x+2))


    ans =


    x^2 + 3*x + 2

fx >> |
```

Fig. 2.25 Expansion of an algebraic expression

You can use the factor command to obtain an array of factors. For instance, according to Fig. 2.26, $x^2 + 3x + 2$ equals to $(x + 2) \times (x + 1)$.

```
Command Window                          ⌄

    >> syms x
    >> factor(x^2+3*x+2)


    ans =


    [x + 2, x + 1]

fx >>
```

Fig. 2.26 Conversion of an algebraic expression to an array of factors

Exercises

1. Use MATLAB to calculate:
   (a) $\lim_{x \to 0} \frac{\sin(2x)}{x} = ?$
   (b) $\frac{d^2}{dx^2} \left(\frac{x^3 - x}{x^2 + 8} \right) = ?$
   (c) $\frac{\partial^2}{\partial x^2} \left(x^3 + 3y^2 + 4xy \right) = ?$
   (d) $\int \cos(x)^2 dx = ?$
   (e) $\int_0^2 \frac{x^3}{x^2 + 64} dx = ?$

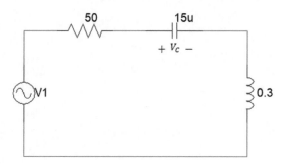

Fig. 2.27 Circuit of exercise 2

2. Use the dsolve command to find the current of circuit shown in Fig. 2.27. $V_1 = 10 + 25\sin(2\pi \times 60t)$. Initial conditions are $V_C = 10$ V and $i_L = 0$ A.
3. Use MATLAB to write the first 8 terms of Taylor series of $x^2 e^{-x}$ at $x = 1$.
4. Use MATLAB to find the Laplace transform of $e^{-t}\sin(6t)$.

Reference for Further Study

1. Symbolic Math Toolbox User's Guide, http://cda.psych.uiuc.edu/matlab_pdf/symbolic_tb.pdf

Numerical Integration and Derivation

3

3.1 Introduction

In the previous chapter you learned how to do symbolic computations in MATLAB environment. In this chapter you will learn how to calculate an integral or derivative with the aid of numeric techniques.

3.2 Trapezoidal Rule

You can use the trapz command in order to calculate an integral using the trapezoidal method. Let's make a numeric data (Fig. 3.1).

Fig. 3.1 Making a simple data

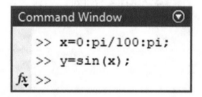

Use the trapz command in order to calculate the integral using the trapezoidal technique (Fig. 3.2). Obtained result is quite close to the correct value (Correct value is $\int_0^\pi \sin(x)dx = 2$).

© The Author(s), under exclusive license to Springer Nature Switzerland AG 2023   73
F. Asadi, *Applied Numerical Analysis with MATLAB®/Simulink®*,
Synthesis Lectures on Engineering, Science, and Technology,
https://doi.org/10.1007/978-3-031-19366-8_3

Fig. 3.2 Numeric integral is
computed with the trapz
command

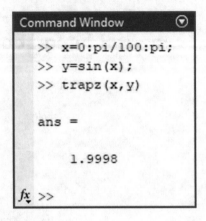

3.3   Simpson's 1/3 Rule

Let's write a code for calculation of $\int_0^\pi \sin(x)dx$ using the Simpson 1/3 rule. Use the edit
command (Fig. 3.3) to run the editor (Fig. 3.4).

Fig. 3.3 edit command

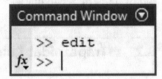

Fig. 3.4 MATLAB editor

Type the following code in MATLAB Editor (Fig. 3.5):

```
Editor - D:\00Numerical Analysis with MATLAB\Programs\                        ⊙ ×
        untitled.m          ✕  +
1       clc
2       clear all
3
4       y=@(x) sin(x); %defines the integrand
5
6       n=6;
7       a=0;
8       b=pi;
9       h=(b-a)/n;
10
11      S=h/3*(y(a)+4*sum(y(a+[1:2:n-1]*h))+2*sum(y(a+[2:2:n-2]*h))+y(b));
12
13      disp("Value of integral using Simpson's 1/3 rule is:")
14      disp(S)
15
```

Fig. 3.5 Code is entered into MATLAB Editor environment

```
clc
clear all

y=@(x) sin(x); % defines the integrand

n=6;                % number of points
a=0;                % bounds
b=pi;               % bounds
h=(b-a)/n;

S=h/3*(y(a)+4*sum(y(a+[1:2:n-1]*h))+2*sum(y(a+[2:2:n-2]*h))+y(b));

disp("Value of integral using Simpson's 1/3 rule is:")
disp(S)
```

After typing the code, press the Ctrl+s or click the icon shown in Fig. 3.6 to save the file. Use the name "Simpson_OneThird_Rule" to save the file (Fig. 3.7).

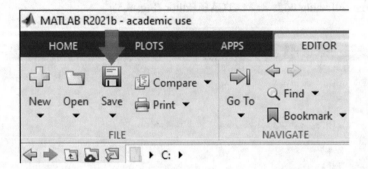

Fig. 3.6   Save icon

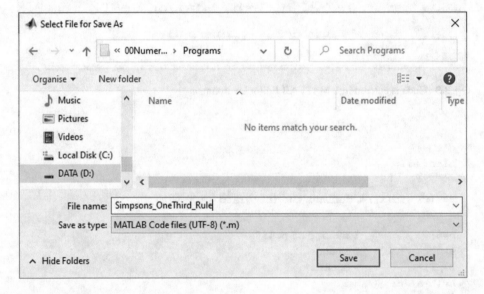

Fig. 3.7   Select file for save as window

After saving the file, press the F5 key of your keyboard or click the icon shown in Fig. 3.8 to run the code. Output of the code is shown in Fig. 3.9. Obtained result is quite close to the correct value which is $\int_0^\pi \sin(x)dx = 2$.

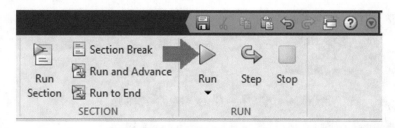

Fig. 3.8   Run icon

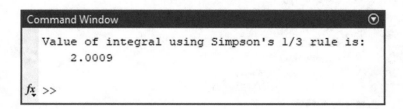

Fig. 3.9   Output of the code

3.4   Simpson 3/8 Rule

In this section we will write a code for calculation of $\int_0^\pi \sin(x)dx$ using the Simpson 3/8 rule. Type the following code in the MATLAB editor (Fig. 3.10).

```
clc
clear all

y=@(x) sin(x); %defines the integrand

n=6;            % n is multiple of three. n is number of points
a=0;            % bounds
b=pi;           % bounds
h=(b-a)/n;

S=y(a)+y(b);
for i=1:n-1
   x=a+i*h;
   if mod(i,3)==0
        S=S+2*y(x);
   else
        S=S+3*y(x);
```

```
Editor - D:\00Numerical Analysis with MATLAB\Programs\Simpsons_Second_Rule.m   ⊙ ✕
  Simpsons_Second_Rule.m  ✕  +
  1         clc
  2         clear all
  3
  4         y=@(x) sin(x); % defines the integrand
  5
  6         n=6;            % n is multiple of three
  7         a=0;
  8         b=pi;
  9         h=(b-a)/n;
 10
 11         S=y(a)+y(b);
 12         for i=1:n-1
 13             x=a+i*h;
 14             if mod(i,3)==0
 15                     S=S+2*y(x);
 16             else
 17                     S=S+3*y(x);
 18             end
 19         end
 20         S=3/8*h*S;
 21
 22         disp("Value of integral using Simpson's 3/8 rule is:")
 23         disp(S)
 24
```

Fig. 3.10 Code is entered into MATLAB Editor environment

```
    end
end
S=3/8*h*S;

disp("Value of integral using Simpson's 3/8 rule is:")
disp(S)
```

After saving the file, press the F5 key of your keyboard or click the icon shown in Fig. 3.8 to run the code. Output of the code is shown in Fig. 3.11. Obtained result is quite close to the correct value which is $\int_0^\pi \sin(x)dx = 2$.

```
Command Window                                          ⊙

  Value of integral using Simpson's 3/8 rule is:
      2.0020

fx >>
```

Fig. 3.11 Output of the code

3.5   Double Integrals

In this section we want to calculate the $\int_0^{1-x} \int_0^1 f(x,y)dxdy$ where $f(x,y) = \frac{1}{\sqrt{x+y}(1+x+y)^2}$. The commands shown in Fig. 3.12 calculate this integral for us.

```
Command Window                                          ⊙

  >> f = @(x,y) 1./(sqrt(x+y).*(1+x+y).^2);
  >> ymax=@(x) 1-x;
  >> integral2(f,0,1,0,ymax)

  ans =

      0.2854

fx >> |
```

Fig. 3.12 Calculation of $\int_0^{1-x} \int_0^1 \frac{1}{\sqrt{x+y}(1+x+y)^2} dxdy$ double integral

3.6   Triple Integrals

In this example section we want to calculate $\int_{-1}^1 \int_0^1 \int_0^\pi f(x,y,z)dxdydz$ where $f(x,y,z) = y\sin(x) + z\cos(x)$. The commands shown in Fig. 3.13 calculate this integral for us.

```
Command Window                                    ⌄
   >> f= @(x,y,z) y.*sin(x)+z.*cos(x);
   >> integral3(f,0,pi,0,1,-1,1)

   ans =

          2.0000

fx >>
```

Fig. 3.13 Calculation of $\int_{-1}^{1}\int_{0}^{1}\int_{0}^{\pi} y sin(x) + z\cos(x)dxdydz$ triple integral

Let's study another example we want to calculate $\int_{-100}^{0}\int_{-100}^{0}\int_{-\infty}^{0} f(x,y,z)dxdydz$ where $f(x,y,z) = \frac{10}{x^2+y^2+z^2+a}$. Commands shown in Fig. 3.14 calculates this integral.

```
Command Window                                    ⌄
   >> a = 2;
   >> f = @(x,y,z) 10./(x.^2+y.^2+z.^2+a);
   >> integral3(f,-Inf,0,-100,0,-100,0)

   ans =

          2.7342e+03

fx >>
```

Fig. 3.14 Calculation of $\int_{-1}^{1}\int_{0}^{1}\int_{0}^{\pi} y\frac{10}{x^2+y^2+z^2+a}dxdydz$ triple integral $(a=2)$

If you need a more accurate result, you can use the format long command (Fig. 3.15).

```
Command Window                                    ⌄
   >> a = 2;
   >> f = @(x,y,z) 10./(x.^2+y.^2+z.^2+a);
   >> format long
   >> integral3(f,-Inf,0,-100,0,-100,0)

   ans =

          2.734244598320928e+03

fx >> |
```

Fig. 3.15 Increasing the accuracy of output

3.7 Derivative

Different discrete formulas exist for derivative. In this section we use the following formulas for first and second order derivative, respectively:

$$f'(x) = \frac{f(x + \Delta x) - f(x - \Delta x)}{2\Delta x}$$

$$f''(x) = \frac{f(x + \Delta x) - 2f(x) + f(x - \Delta x)}{\Delta x^2} \quad (3.1)$$

Following code computes the first and second derivative of $4x^2 + 5x^3$ at the point that entered by the user. Enter the code to the MATLAB editor and save it with the name Derivative.m (Fig. 3.16).

```
1    clc
2    clear all
3
4    f=@(x) 4*x^2+5*x^3;
5
6    x0=input("please enter the point(x0): ");
7    h=input("please enter the h: ");
8
9    disp("First order derivative: ")
10   disp((f(x0+h)-f(x0-h))/(2*h))
11
12   disp("Second order derivative: ")
13   disp((f(x0+h)-2*f(x0)+f(x0-h))/(h^2))
14
```

Fig. 3.16 Code is entered into MATLAB Editor environment

```
clc
clear all

f=@(x) 4*x^2+5*x^3;

x0=input("please enter the point(x0): ");
h=input("please enter the h: ");

disp("First order derivative: ")
```

```
disp((f(x0+h)-f(x0-h))/(2*h))

disp("Second order derivative: ")
disp((f(x0+h)-2*f(x0)+f(x0-h))/(h^2))
```

Sample run of the code in Fig. 3.16 is shown in Fig. 3.17.

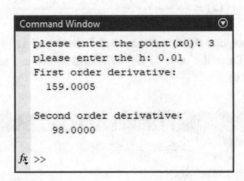

Fig. 3.17 Output of the code

Let's calculate the second order derivative of $\sin(2x)$ for $[0, 2\pi]$ interval. Following code computes the second order derivative and draws its graph.

```
clc
clear all

f=@(x) sin(2*x);
n=1000;
h=2*pi/n;
x=0:h:2*pi;
y=f(x);

d2f=(y(3:n)-2*y(2:n-1)+y(1:n-2))/(h^2);
plot(x,y,'b');
hold on
plot(x(2:n-1),d2f,'r');
grid on
```

Enter the code to MATLAB Editor (Fig. 3.18) and run it. Output of entered code is shown in Fig. 3.19.

```
Editor - D:\00Numerical Analysis with MATLAB\Programs\SecondOrderDerivativeInterval.m ⊙ ×
SecondOrderDerivativeInterval.m  ✕  +
1       clc
2       clear all
3
4       f=@(x) sin(2*x);
5       n=1000;
6       h=2*pi/n;
7       x=0:h:2*pi;
8       y=f(x);
9
10      d2f=(y(3:n)-2*y(2:n-1)+y(1:n-2))/(h^2);
11      plot(x,y,'b');
12      hold on
13      plot(x(2:n-1),d2f,'r');
14      grid on
```

Fig. 3.18 Code is entered into MATLAB Editor environment

Fig. 3.19 Output of the code

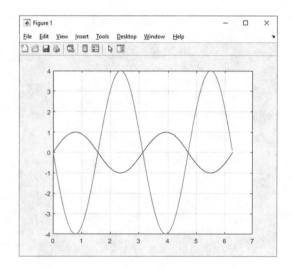

Exercises

1. Use the Simpson rules to calculate the $\int_0^1 e^{-x} \sin(3x)dx$ (use the 0.01 steps).
2. Use the trapezoidal rules to calculate the $\int_0^1 e^{-x} \sin(3x)dx$ (use the 0.01 steps).
3. Use MATLAB to calculate the following integrals.
   (a) $\int_0^{1-x} \int_0^1 \frac{1}{\sqrt{x+4y}(1+y)^2}dxdy$

(b) $\int_{-1}^{1} \int_{0}^{1} \int_{0}^{\pi} y^2 \sin(2x) + z^2 \cos(x^2 + y^2) dx dy dz$

4. Use formula (3.1) to calculate first and second derivative of $e^{-x} \sin(3x)$ at $x = 5$.

References for Further Study

1. Applied Numerical Methods with MATLAB for Engineers and Scientists, Chapra S., Mc Graw Hill, 2005.
2. Numerical Analysis, Burden, RL., Fairs D., Burden AM., Cengage Learning, 2015.
3. Numerical Methods: Using MATLAB, Lindfield G., Penny J., Academic press, 2018.

Statistics with MATLAB®

4

4.1   Introduction

In this chapter you will learn how to do statistical calculations in MATLAB.

4.2   Sum of Elements

You can use the sum command to obtain the sum of elements of a vector. An example is given in Fig. 4.1.

Fig. 4.1 sum command

```
Command Window                              ⊙

>> x=[5 6 8 9 10 11 14];
>> sum(x)

ans =

    63

fx >> |
```

© The Author(s), under exclusive license to Springer Nature Switzerland AG 2023   85
F. Asadi, *Applied Numerical Analysis with MATLAB®/Simulink®*,
Synthesis Lectures on Engineering, Science, and Technology,
https://doi.org/10.1007/978-3-031-19366-8_4

You can use the sum command to obtain the summation of elements of a column of a matrix. An example is given in Fig. 4.2.

Fig. 4.2 In matrices, the sum command returns the summation of columns

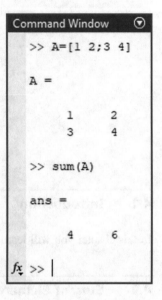

4.3   Average

You can calculate the average of a vector with the aid of mean function. An example is given in Fig. 4.3.

Fig. 4.3 mean command

You can calculate the average of a vector with the aid of commands shown in Fig. 4.4 as well. The result obtained in Fig. 4.4 is the same as the one in Fig. 4.3.

Fig. 4.4 Computation of average

```
Command Window

>> x=[5  6  8  9  10  11  14];
>> sum(x)/length(x)

ans =

        9

fx >> |
```

As another example assume that we want to calculate the average of a students. The student took 60 from physics (3 credits), 70 from mathematics (5 credits), 65 from chemistry (3 credits), 90 from history (3 credits) and 80 from literature (2 credits). The average of the student can be calculated with the aid of commands shown in Fig. 4.5.

Fig. 4.5 Weighted averaging

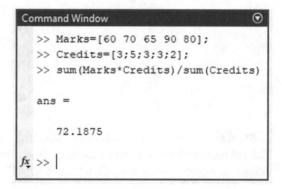

```
Command Window

>> Marks=[60  70  65  90  80];
>> Credits=[3;5;3;3;2];
>> sum(Marks*Credits)/sum(Credits)

ans =

      72.1875

fx >> |
```

The mean function can work on matrices as well. The mean function returns the average of each column (summation of elements of the column divided by the number of rows) when it works on matrices.

4.4   Variance and Standard Deviation

Variance of a vector can be obtained with the aid of var command (Fig. 4.6). The var command takes matrices as well. When the input to var function is a matrix, it returns the variance of each column.

Fig. 4.6 var command

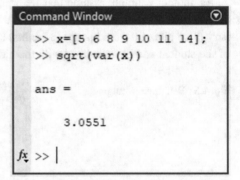

The standard deviation can be calculated by taking the square root of the variance (Fig. 4.7).

Fig. 4.7 Calculation of standard deviation

Standard deviation of a vector can be obtained with the aid of std command (Fig. 4.8). The std command takes matrices as well. When the input to std function is a matrix, it returns the standard deviation of each column.

Fig. 4.8 std command

4.5 Factorial

You can use the factorial function in order to calculate the factorial of an integer number. The extension of the factorial function is gamma function. Remember that $\Gamma(n) = (n-1)!$ when $n \in \mathbb{N}$ (Fig. 4.9).

Fig. 4.9 factorial and gamma commands

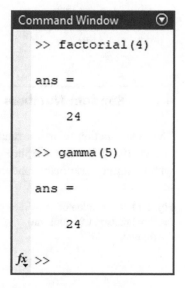

```
>> factorial(4)

ans =

    24

>> gamma(5)

ans =

    24

fx >>
```

4.6 Combination

Number of combinations of n items taken k at a time is equal to $\begin{pmatrix} n \\ k \end{pmatrix} = \frac{n!}{(n-k)!k!}$.

nchoosek function in MATLAB can calculate $\begin{pmatrix} n \\ k \end{pmatrix}$ for us. An example is given in Fig. 4.10.

Fig. 4.10 nchoosek command

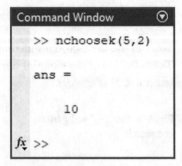

4.7   Random Numbers

The rand function returns a random scalar drawn from the uniform distribution in the interval (0, 1). For instance, in Fig. 4.11 we asked the MATLAB to make a 1×3 matrix with uniformly distributed random elements.

Fig. 4.11 Generation of random numbers with uniform distribution

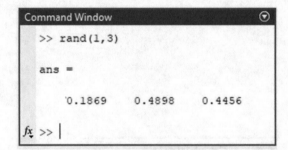

The randn function generates random numbers whose are normally distributed with mean 0, variance $\sigma^2 = 1$ and standard deviation $\sigma = 1$. For instance, in Fig. 4.12 we asked the MATLAB to make a 1×3 matrix with normally distributed random elements.

Fig. 4.12 Generation of random numbers with normal distribution

```
Command Window                                    ⊙
    >> randn(1,3)

    ans =

          0.6277      1.0933      1.1093

fx >>
```

In general, you can generate N random numbers in the interval (a, b) with the formulas r = a + (b − a).*rand(N, 1) or r = a + (b − a).*randn(N, 1).

For instance, the commands in Fig. 4.13 generates two random number selected from the (1, 7) interval.

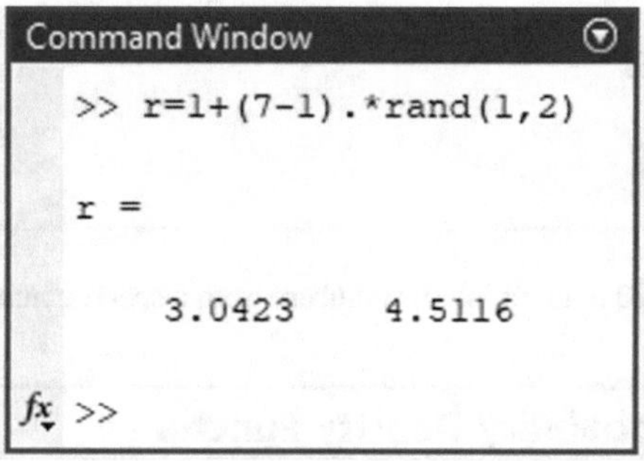

```
Command Window                          ⊙

   >> r=1+(7-1).*rand(1,2)

   r =

           3.0423      4.5116

fx >>
```

Fig. 4.13 Generation of random numbers from a specific interval

You can use the floor command (Fig. 4.14) in order to get rid of the decimal part of Fig. 4.13. Such a technique is useful when you need random integer numbers. The command shown in Fig. 4.15 generates 20 random integer ({1, 2, 3, 4, 5, 6}) number with uniform distribution.

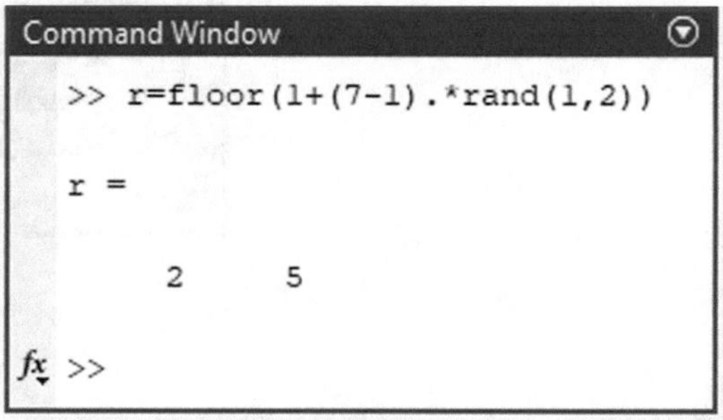

```
Command Window                          ⊙

   >> r=floor(1+(7-1).*rand(1,2))

   r =

              2       5

fx >>
```

Fig. 4.14 Generation of 2 random integer numbers from a specific interval

```
Command Window                                                              ⊙

>> r=floor(1+(7-1).*rand(1,20))

r =

  Columns 1 through 15

     5     4     4     6     2     5     5     3     4     1     1     4     5     6     1

  Columns 16 through 20

     4     3     1     3     1

fx >> |
```

Fig. 4.15 Generation of 20 random integer numbers from a specific interval

4.8   Normal Probability Density Function

The normal probability density function (pdf) is $y = f(x|\mu, \sigma) = \frac{1}{\sigma\sqrt{2\pi}}e^{\frac{-(x-\mu)^2}{2\sigma^2}}$ for $x \in \mathbb{R}$. MATLAB function Y = normpdf (X, MU, SIGMA) returns the pdf of the normal distribution with mean MU and standard deviation SIGMA, evaluated at the values in X. For instance, the code in Fig. 4.16 computes the pdf of a standard normal distribution, with parameters μ equal to 0 and σ equal to 1 and draws its graph (Fig. 4.17).

Fig. 4.16   normpdf command

```
Command Window                        ⊙

  >> x=[-3:0.1:3];
  >> y=normpdf(x,0,1);
  >> plot(x,y)
fx >>
```

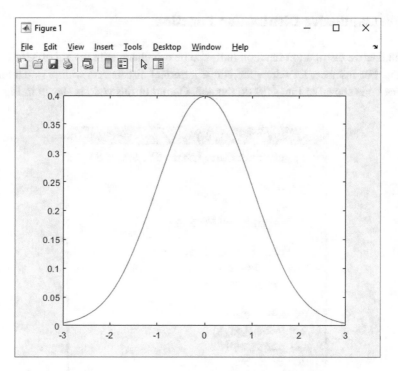

Fig. 4.17   Normal distribution with average of 0 and standard deviation of 1

The graph shown in Fig. 4.17 can be generated with the aid of the code shown in Fig. 4.18.

```
Command Window
>> x=[-3:0.1:3];
>> sigma=1;mu=0;
>> Y=1/sigma/sqrt(2*pi)*exp(-(x-mu).^2/2/sigma^2);
>> plot(x,y);
fx >>
```

Fig. 4.18   Another way to generate the graph shown in Fig. 4.17

4.9   Cumulative Distribution Function

The cumulative distribution function (cdf) can be calculated with the aid of MATLAB cdf function. For instance, let's draw the cdf of a normal distribution with $\mu = 0.8$ and $\sigma = 0.6$. The code shown in Fig. 4.19 do this job. Output of this code is shown in Fig. 4.20.

```
Command Window                                    ⊙
 >> pd=makedist('Normal',0.8,0.6)

 pd =

   NormalDistribution

   Normal distribution
         mu = 0.8
      sigma = 0.6

 >> x=-3:0.1:3;
 >> y=cdf(pd,x);
 >> plot(x,y)
 >> grid on
fx >> |
```

Fig. 4.19 cfd function

Fig. 4.20 Output of the code in Fig. 4.19

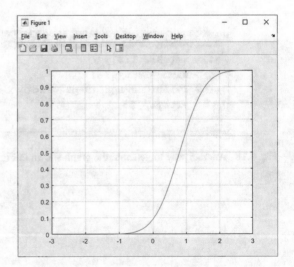

You can click on any point of the graph in order to read its value. For instance, value of cdf at $x = 0.3$ is 0.202328 (Fig. 4.21). This means that $\int_{-\infty}^{0.3} \frac{1}{\sigma\sqrt{2\pi}} e^{\frac{-(x-\mu)^2}{2\sigma^2}} dx = 0.202328$. In this example $\sigma = 0.6$ and $\mu = 0.8$.

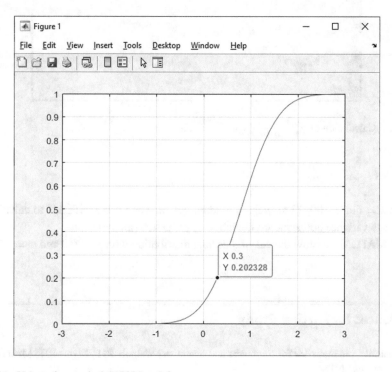

Fig. 4.21 Value of curve is 0.202328 at 0.3

The commands shown in Fig. 4.22 calculates the $\int_{-20}^{0.3} \frac{1}{\sigma\sqrt{2\pi}} e^{\frac{-(x-\mu)^2}{2\sigma^2}} dx$. As you see the result is quite close to the value shown in Fig. 4.21. You can decrease the lower bound of the integral and obtain more accurate results.

```
Command Window                                                    ▼

>> sigma=0.6;mu=0.8;
>> x=-20:0.01:0.3;
>> Y=1/sigma/sqrt(2*pi)*exp(-(x-mu).^2/2/sigma^2);
>> trapz(x,Y)

ans =

    0.2023

fx >>
```

Fig. 4.22 Calculation of $\int_{-20}^{0.3} \frac{1}{0.6\times\sqrt{2\pi}} e^{\frac{-(x-0.8)^2}{2\times0.36}} dx$

Exercises

1. Use x = randn(100, 1) to make a random vector. Then use MATLAB to calculate the average variance and standard deviation of generated random vector.
2. Use MATLAB to draw the cdf of a normal distribution with $\mu = 0.4$ and $\sigma = 0.3$.

References for Further Study

1. Probability and Statistics for Engineers and Scientists, Walpole RH., Myers RE., Myers SL., Ye KE, Pearson, 2013.
2. Statistics for Engineers and Scientists, Navidi W., McGraw Hill, 2019.

Impulse and Step Response of Linear Systems

5

5.1   Introduction

In this chapter you will learn how to draw the response of linear systems to Dirac delta (unit impulse) function and Heaviside (unit) step functions. Dirac delta function ($\delta(t)$) is a function whose value is zero everywhere except at zero, and whose integral over the entire real line is equal to one. The step function ($H(t)$) is zero for negative arguments and one for positive arguments (Fig. 5.1).

Fig. 5.1 a Unit impulse function; **b** unit step function

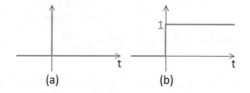

(a)   (b)

For instance, assume that we want to obtain the impulse response of $\frac{d^2 y(t)}{dt} + 6\frac{dy(t)}{dt} + 100y(t) = 100u(t)$. Note that $y(t)$ and $u(t)$ show the system output and input, respectively. We can use the Laplace transform to obtain the impulse response of given system: $s^2 Y(s) + 6s Y(s) + 100Y(s) = 100U(s)$. When $u(t) = \delta(t)$, $U(s) = 1$, so $s^2 Y(s) + 6s Y(s) + 100Y(s) = 100$. This means $Y(s) = \frac{100}{s^2+6s+100}$. Therefor inverse Laplace of $\frac{100}{s^2+6s+100}$ gives the impulse response of the system.

Let's see how we can obtain the step response of $\frac{d^2 y(t)}{dt} + 6\frac{dy(t)}{dt} + 100y(t) = 100u(t)$ as well. We can use the Laplace transform to obtain the step response of given system: $s^2 Y(s) + 6s Y(s) + 100Y(s) = 100U(s)$. When $u(t) = H(t)$, $U(s) = \frac{1}{s}$, so $s^2 Y(s) + 6s Y(s) + 100Y(s) = \frac{100}{s}$. This means $Y(s) = \frac{100}{s(s^2+6s+100)}$. Therefor inverse Laplace of $\frac{100}{s(s^2+6s+100)}$ gives the step response of the system.

© The Author(s), under exclusive license to Springer Nature Switzerland AG 2023   97
F. Asadi, *Applied Numerical Analysis with MATLAB®/Simulink®*,
Synthesis Lectures on Engineering, Science, and Technology,
https://doi.org/10.1007/978-3-031-19366-8_5

5.2   Impulse Response of Dynamical Systems

Impulse and step response of dynamical systems can be drawn easily with MATLAB. Assume we want to draw the impulse and step responses of $\frac{d^2 y(t)}{dt} + 6\frac{dy(t)}{dt} + 100y(t) = 100u(t)$. Note that $y(t)$ and $u(t)$ show the system output and input, respectively. Transfer function of the system is defined as Laplace transform of the output divided by Laplace transform of the input, i.e. $\frac{\mathcal{L}(y(t))}{\mathcal{L}(u(t))}$. We use the $H(s)$ to show the transfer function of the system. For the given system $H(s) = \frac{\mathcal{L}(y(t))}{L(u(t))} = \frac{100}{s^2+6s+100}$. The commands shown in Figs. 5.2 or 5.3 defines the given system.

```
Command Window                        ⊙
   >> H=tf([100],[1 6 100]);
fx >> |
```

Fig. 5.2 Defining the H(s) $= \frac{100}{s+6s+100}$

```
Command Window                        ⊙
   >> s=tf('s');
   >> H=100/(s^2+6*s+100);
fx >> |
```

Fig. 5.3 Defining the H(s) $= \frac{100}{s+6s+100}$

The impulse command (Fig. 5.4) draws the impulse response of the system (Fig. 5.5).

```
Command Window                        ⊙
   >> H=tf([100],[1 6 100]);
   >> impulse(H), grid on
fx >>
```

Fig. 5.4 Impulse response of H(s) $= \frac{100}{s+6s+100}$

Fig. 5.5 Output of code in
Fig. 5.4

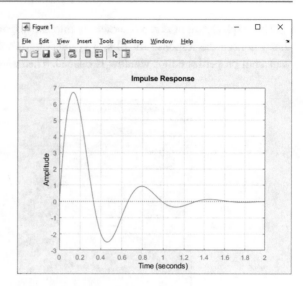

You can add cursors to the graph by clicking on the graph (Fig. 5.6). You can remove the added cursor by right clicking on it and click the delete.

Fig. 5.6 Addition of cursor to
the graph

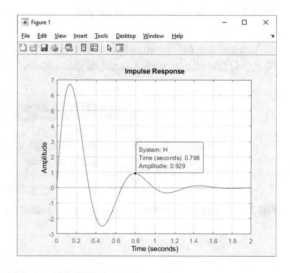

The graph shown in Fig. 5.6 shows the impulse response for [0 s, 2 s] interval. Assume that you need the response for [0 s, 3 s] interval. Simply double click on the graph and wait until properties editor window is opened. Then enter the desired range (Fig. 5.7).

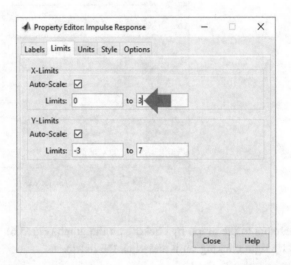

Fig. 5.7 Determining the limits of the plot

You can use the command shown in Figs. 5.8 or 5.9 to draw the impulse response for [0 s, 3 s] interval, as well. Output of these codes are shown in Fig. 5.10.

```
Command Window
>> s=tf('s');
>> H=tf(100,[1 6 100]);
>> impulse(H,3)
fx >>
```

Fig. 5.8 Drawing the impulse response for the desired time interval

```
Command Window
>> time=[0:0.01:3];
>> impulse(H,time), grid on
fx >>
```

Fig. 5.9 Drawing the impulse response for the desired time interval

Fig. 5.10 Output of code in
Figs. 5.8 and 5.9

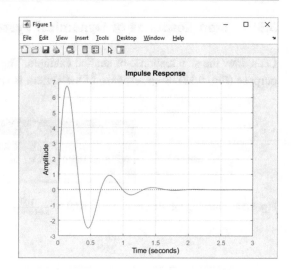

You can measure the impulse response characteristics by right clicking on the graph
and selecting the characteristics (Fig. 5.11).

Fig. 5.11 Measuring the
characteristics of the graph

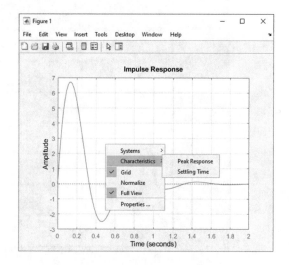

5.3   Step Response of Dynamical Systems

Let's draw the step response of studied example. The step command can be used for this purpose (Fig. 5.12). Step response of the system is shown in Fig. 5.13.

```
Command Window                        ⦿
  >> H=tf([100],[1 6 100]);
  >> step(H), grid on
fx >>
```

Fig. 5.12 Drawing the step response of $H(s) = \frac{100}{s+6s+100}$

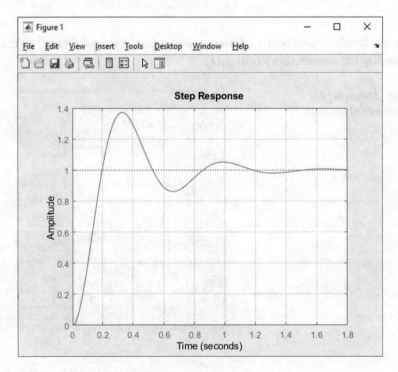

Fig. 5.13 Output of code shown in Fig. 5.12

Exercises

1. Assume that $\frac{d^2y(t)}{dt} + 6\frac{dy(t)}{dt} + y(t) = \frac{du(t)}{dt} + 6u(t)$. $y(t)$ and $u(t)$ show the system output and input, respectively.

   (a) Use hand analysis to obtain the system transfer function.
   (b) Use hand analysis to obtain the impulse response of the system.
   (c) Use hand analysis to obtain the step response of the system.
   (d) Use MATLAB to draw the impulse response of the system.
   (e) Use MATLAB to draw the step response of the system.
   (f) Compare result of hand analysis with MATLAB result.

Reference for Further Study

1. Signal and Systems, Oppenheim A., Willsky A., Nawab H., Pearson, 1996.

Solving Differential Equations in Simulink®

6

6.1 Introduction

In this chapter you will learn the basics of Simulink environment. Simulink is a graphical simulation environment which accompany MATLAB and you can make the model of systems with the aid of ready to use blocks. Thanks to many ready to use blocks, there is a little or no need to write code (program) when you are working in Simulink environment. This property makes the Simulink an easy tool for many branches of science and engineering.

6.2 Solving a Linear Differential Equation in Simulink

In this section we want to solve the following differential equation problem:

$$\ddot{y}(t) + 5\dot{y}(t) + 3y(t) = u(t), \, y(0) = 5, \, \dot{y}(0) = 2 \tag{6.1}$$

$y(t)$ and $u(t)$ show the output and input, respectively. $u(t)$ is given as:

$$u(t) = 3 + 0.7 \sin\left(0.5t + \frac{\pi}{4}\right) \tag{6.2}$$

First of all, we need to convert the given differential equation into state space. We define two new variables:

$$\begin{cases} x_1(t) = y(t) \\ x_2(t) = \dot{y}(t) \end{cases} \tag{6.3}$$

The given system can be written as:

© The Author(s), under exclusive license to Springer Nature Switzerland AG 2023
F. Asadi, *Applied Numerical Analysis with MATLAB®/Simulink®*,
Synthesis Lectures on Engineering, Science, and Technology,
https://doi.org/10.1007/978-3-031-19366-8_6

$$\begin{cases} \dot{x}_1(t) = x_2(t) \\ \dot{x}_2(t) = -3x_1(t) - 5x_2(t) + u(t) \end{cases}, \quad \begin{bmatrix} x_1(0) \\ x_2(0) \end{bmatrix} = \begin{bmatrix} 5 \\ 2 \end{bmatrix} \quad (6.4)$$

Note that for an nth order differential equation, you need n state variables:

$$\begin{cases} x_1(t) = y(t) \\ x_2(t) = \dot{y}(t) \\ x_3(t) = \ddot{y}(t) \\ \quad \vdots \\ x_n(t) = y(t)^{(n-1)} \end{cases} \quad (6.5)$$

Enter into the Simulink environment (Fig. 6.1) and select a blank model (Fig. 6.2). This opens the Simulink environment (Fig. 6.3).

Fig. 6.1 Simulink command

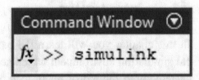

Fig. 6.2 Simulink start page

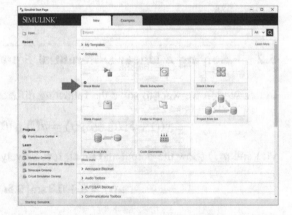

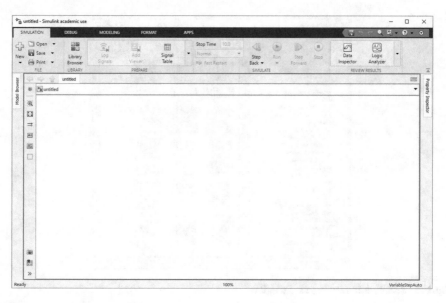

Fig. 6.3 Simulink environment

Click on the library browser (Fig. 6.4). This opens the Simulink library browser window shown in Fig. 6.5.

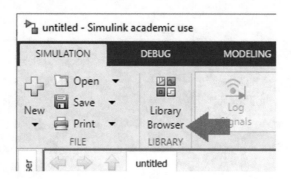

Fig. 6.4 Library browser icon

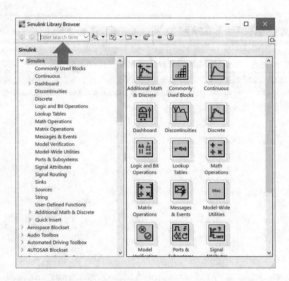

Fig. 6.5 Simulink library browser window

Go to the continuous section and drag and drop two integrator block (Fig. 6.6) to the working area (Fig. 6.7). You can use the "Enter search term" box (Fig. 6.5) in order to search for a block. For instance, you can search for integrator if you don't know or forget its location.

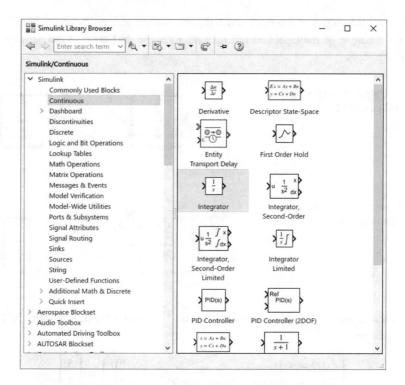

Fig. 6.6 Integrator block

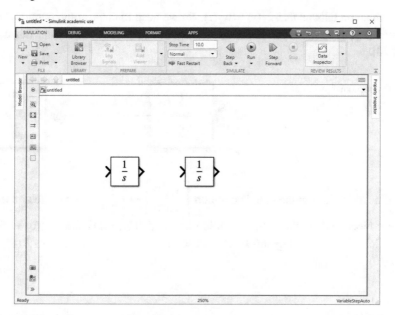

Fig. 6.7 Addition of two integrator block to the working area

Let's call the output of the right and left integrator x_1 and x_2, respectively (Fig. 6.8). Therefor the input of the right and left integrator are the derivative of the corresponding output (Fig. 6.9).

Fig. 6.8 Output of integrators are named x1 and x2

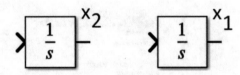

Fig. 6.9 Input of integrators are derivative of corresponding output

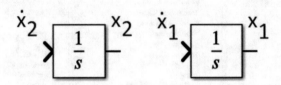

It's time to implement the state space Eq. 6.4. The equation is re-written below for ease of reference:

$$\begin{cases} \dot{x}_1(t) = x_2(t) \\ \dot{x}_2(t) = -3x_1(t) - 5x_2(t) + u(t) \end{cases}, \quad \begin{bmatrix} x_1(0) \\ x_2(0) \end{bmatrix} = \begin{bmatrix} 5 \\ 2 \end{bmatrix} \tag{6.4}$$

First equation tells us that $\dot{x}_1(t) = x_2(t)$. So we need to connect the output of the left integrator $(x_2(t))$ to the input of the right integrator $(\dot{x}_1(t))$ to implement this equation (Fig. 6.10).

Fig. 6.10 Implementation of $\dot{x}_1(t) = x_2(t)$

It's time to implement the initial condition $\begin{bmatrix} x_1(0) \\ x_2(0) \end{bmatrix} = \begin{bmatrix} 5 \\ 2 \end{bmatrix}$. Double click on the right integrator block and set the initial condition box to 5 (Fig. 6.11) since $x_1(0) = 5$. Then double click on the left integrator block and set the initial condition box to 2 (Fig. 6.12) since $x_2(0) = 2$.

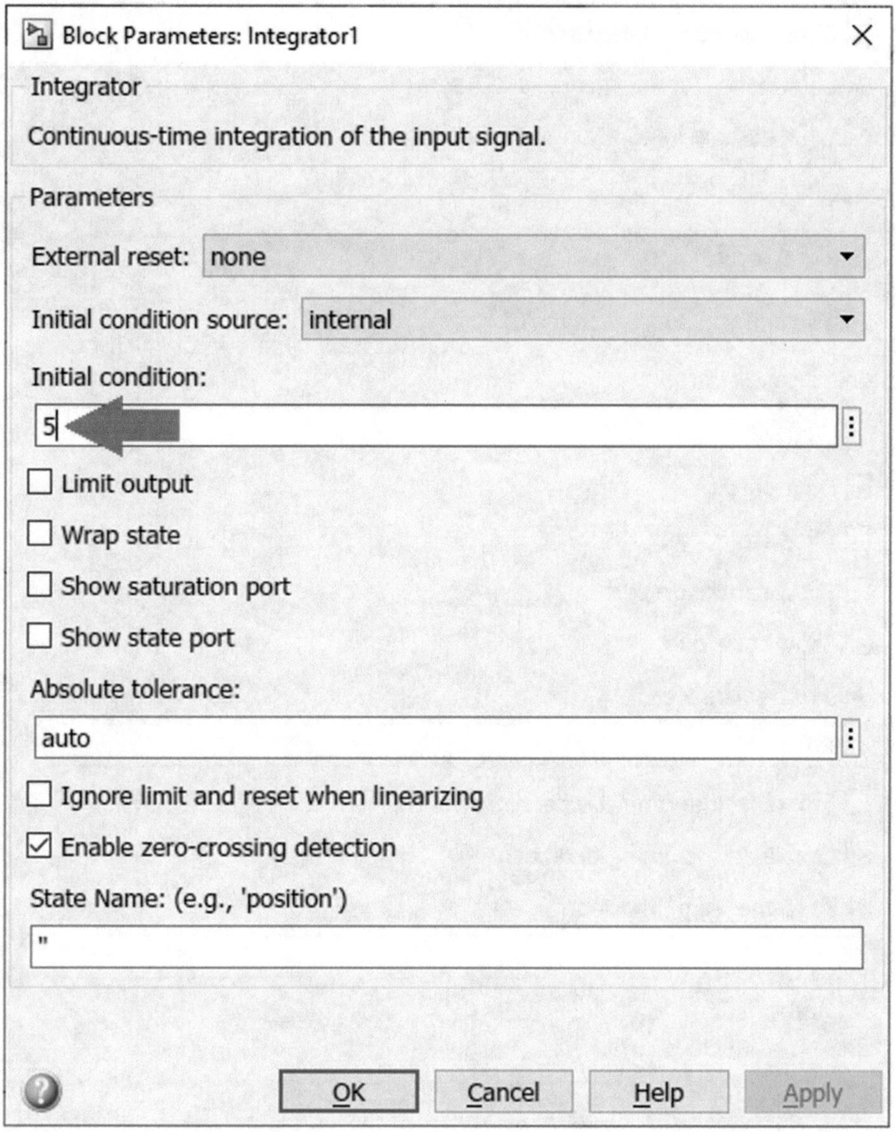

Fig. 6.11 Entering the initial condition to the right integrator

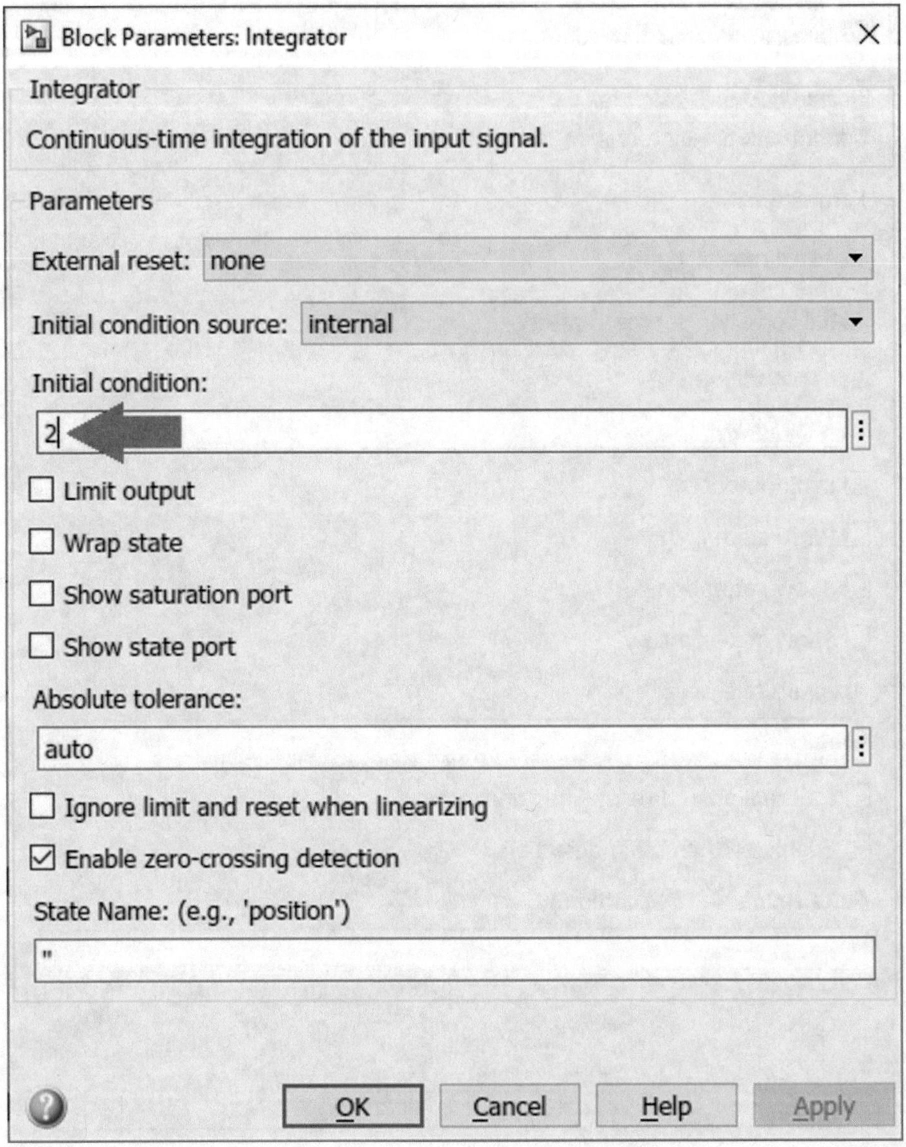

Block Parameters: Integrator   ✕

Integrator

Continuous-time integration of the input signal.

Parameters

External reset: none   ▼

Initial condition source: internal   ▼

Initial condition:

2

☐ Limit output

☐ Wrap state

☐ Show saturation port

☐ Show state port

Absolute tolerance:

auto

☐ Ignore limit and reset when linearizing

☑ Enable zero-crossing detection

State Name: (e.g., 'position')

"

OK   Cancel   Help   Apply

Fig. 6.12 Entering the initial condition to the left integrator

It's time to implement the second equation ($\dot{x}_2(t) = -3x_1(t) - 5x_2(t) + u(t)$). We need to add a sum (Fig. 6.13) and gain blocks (Fig. 6.14) to our model (Fig. 6.15). You can rotate a block by left clicking on it and pressing the Ctrl+R keys of your keyboard.

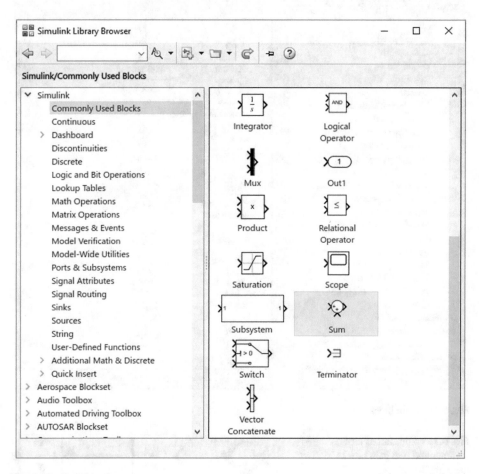

Fig. 6.13 Sum block

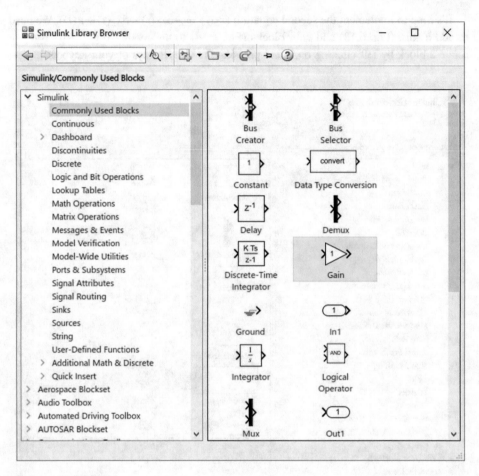

Fig. 6.14 Gain block

Fig. 6.15 Addition of a
summer and two gain blocks to
the Simulink model

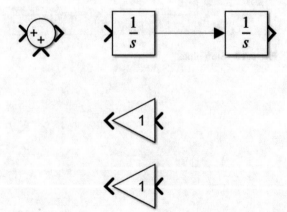

Double click on the summer block and do the settings similar to Fig. 6.16. Now, the summer block has three inputs (Fig. 6.17).

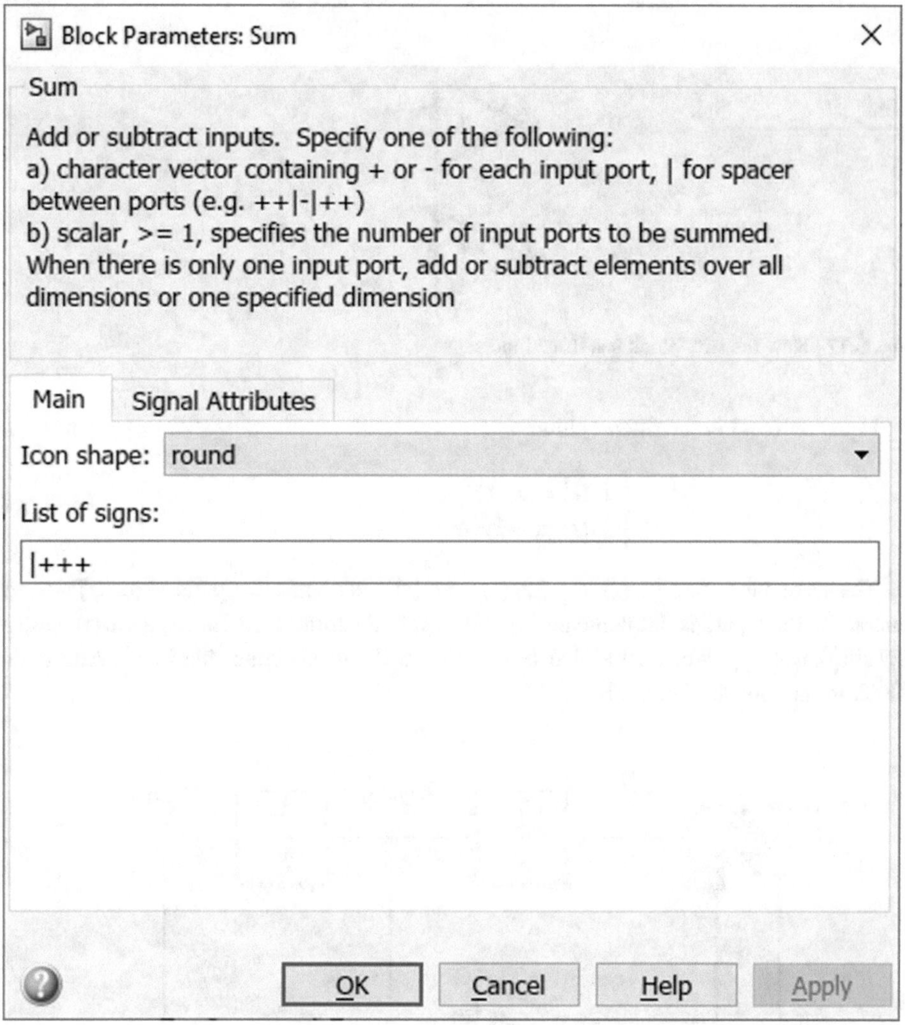

⊡ Block Parameters: Sum ✕

Sum

Add or subtract inputs. Specify one of the following:
a) character vector containing + or - for each input port, | for spacer between ports (e.g. ++|-|++)
b) scalar, >= 1, specifies the number of input ports to be summed. When there is only one input port, add or subtract elements over all dimensions or one specified dimension

---

Main   Signal Attributes

Icon shape: round   ▼

List of signs:

| +++

       [OK]   [Cancel]   [Help]   [Apply]

Fig. 6.16 Settings of the sum block

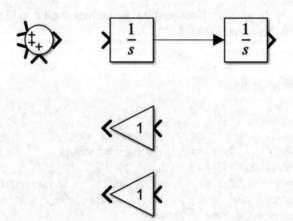

Fig. 6.17 Now the sum block has three inputs

Equation (6.4) is rewritten below:

$$\begin{cases} \dot{x}_1(t) = x_2(t) \\ \dot{x}_2(t) = -3x_1(t) - 5x_2(t) + u(t) \end{cases} \tag{6.4}$$

The second equation ($\dot{x}_2(t) = -3x_1(t) - 5x_2(t) + u(t)$) has two parts: $-3x_1(t) - 5x_2(t)$ and $u(t)$. First part is implemented in Fig. 6.18. According to Eq. (6.2), $u(t) = 3 + 0.7\sin\left(0.5t + \frac{\pi}{4}\right)$. We need a sine block (Fig. 6.19) to generate this input. Add a sine block to the model (Fig. 6.20).

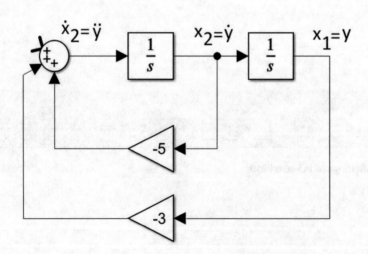

Fig. 6.18 Implementation of $-3x_1(t) - 5x_2(t)$

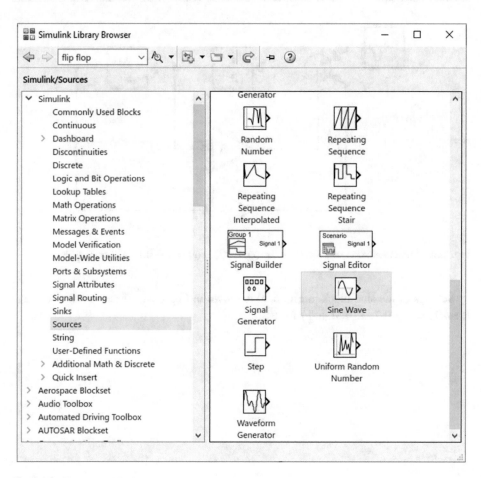

Fig. 6.19 Sine wave block

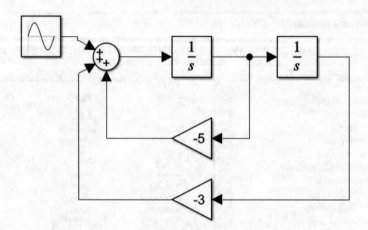

Fig. 6.20 Implementation of $-3x_1(t) - 5x_2(t) + u(t)$. $u(t) = 3 + 0.7\sin\left(0.5t + \frac{\pi}{4}\right)$

Settings of the sine block in Fig. 6.20 is shown in Fig. 6.21. These settings generates the $u(t) = 3 + 0.7\sin\left(0.5t + \frac{\pi}{4}\right)$ for us.

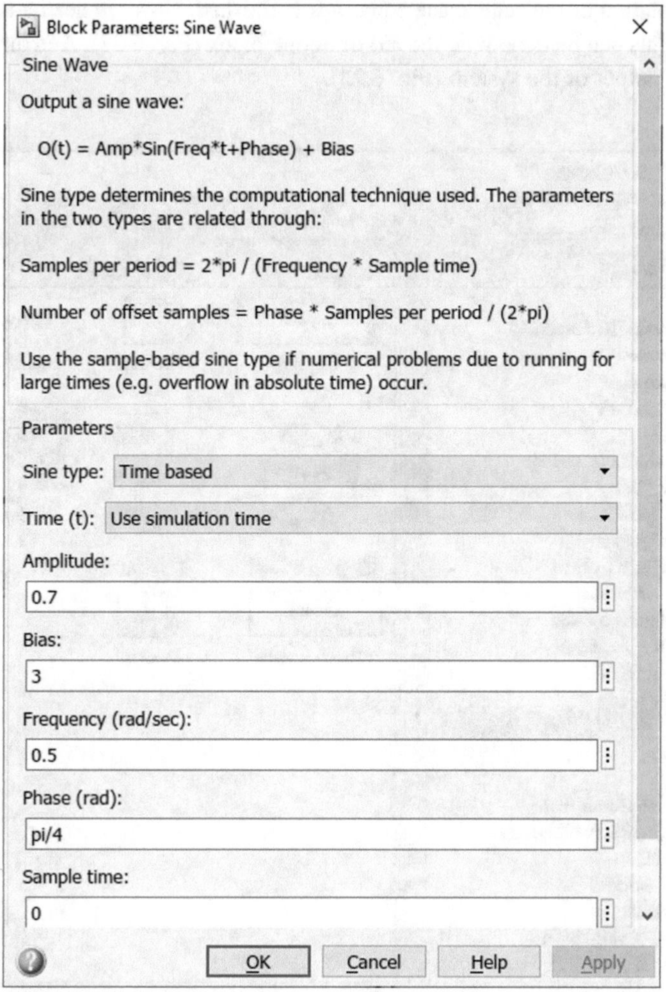

Fig. 6.21   Sine wave block settings

Implementation of the state space equations is finished. Now we need a block to show the output. This can be done with the aid of Scope block (Fig. 6.22). Add an oscilloscope block to the output of the system (Fig. 6.23).

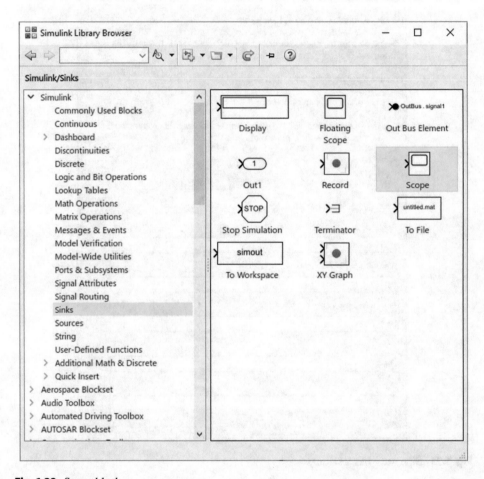

Fig. 6.22 Scope block

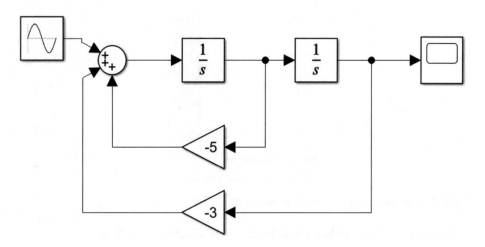

Fig. 6.23 Completed Simulink model

We want to observe the output for [0 s, 10 s] interval. So, you need to enter 10 to Stop Time box (Fig. 6.24). Name of the method that is used to solve the Simulink model is shown in bottom right side of the window (Fig. 6.25). ode45 is based on an explicit Runge–Kutta (4, 5) formula, the Dormand-Prince pair. It is a one-step solver; that is, in computing $y(t_n)$, it needs only the solution at the immediately preceding time point, $y(t_{n-1})$. In general, ode45 is the best solver to apply as a "first try" for most problems. Click the Run icon (Fig. 6.24) in order to run the model and wait until Simulink finish its job.

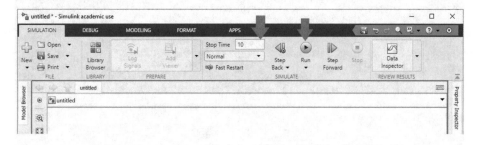

Fig. 6.24 Stop time box and run icon

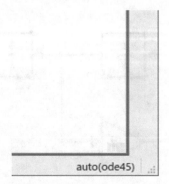

Fig. 6.25 Used solver is shown in the bottom of the screen

Double click the scope block in order to see its waveform (Fig. 6.26).

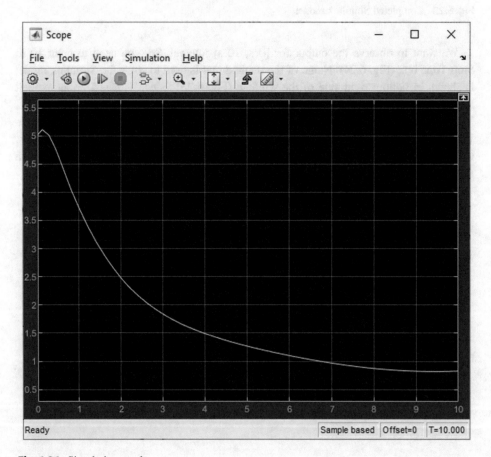

Fig. 6.26 Simulation result

Click on the cursor measurements icon (Fig. 6.27). This adds two cursors to the scope block and permits you to read the values of the waveform easily.

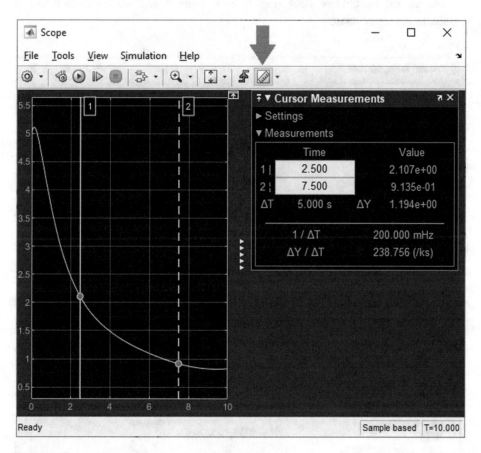

Fig. 6.27 Cursor measurement icon

6.3   Multiplexer Block

You can use the multiplexer block (Fig. 6.28) in order to see two (or more than two) waveforms simultaneously.

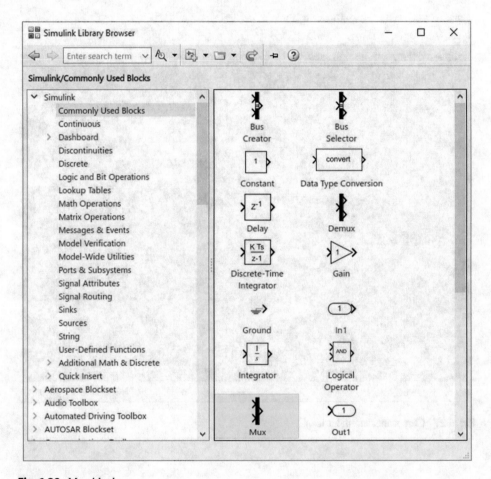

Fig. 6.28 Mux block

Let's observe the output of the integrators simultaneously. Change the model of previous example to what shown in Fig. 6.29. By default, the multiplexer block has two inputs. If you double click on the multiplexer block, you can increase the number of inputs.

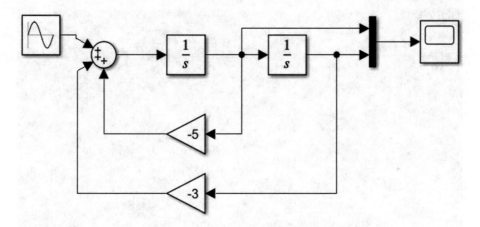

Fig. 6.29 Multiplexer block is used to show two waveforms simultaneously on a scope

Now run the model shown in Fig. 6.29. After running the model, double click the scope block in order to see the waveforms (Fig. 6.30). The scope block shows two waveforms: one with yellow color and the other one with blue color. The yellow one belongs to the signal that enters the upper input of the multiplexer and the blue one belongs to the signal that enters to the lower input of the multiplexer (Fig. 6.31). Default colors for a multiplexer with five inputs are shown in Fig. 6.32.

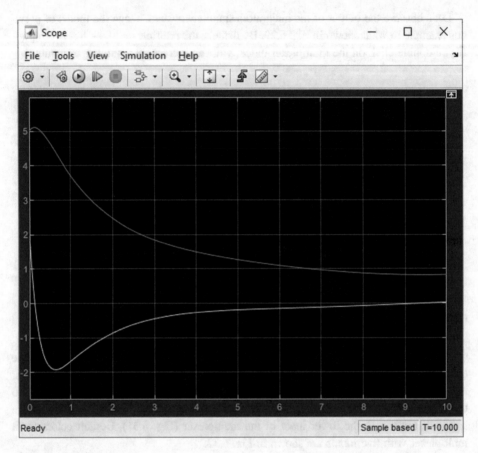

Fig. 6.30 Simulation result

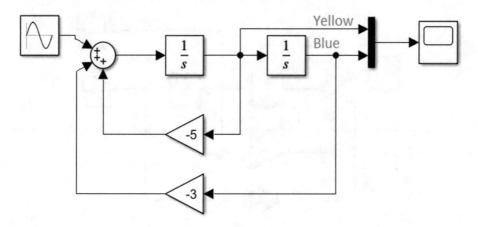

Fig. 6.31 Default colors for a two inputs multiplexer

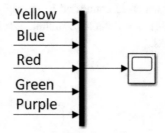

Fig. 6.32 Default colors for a five inputs multiplexer

6.4   Giving Name to Scope Blocks

You can have many scopes in Simulink model. It is a good idea to give meaningful names to scopes which describe the waveform that they show. Giving meaningful names to scope blocks is very useful specially when your model contains many scope blocks. Because it helps you to understand quickly what is what. Let's study an example: Consider the model shown in Fig. 6.33.

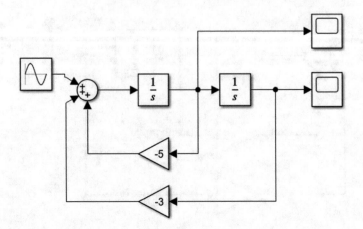

Fig. 6.33 Sample Simulink model

Click on the scope block which is connected to the right integrator. This shows the name of the block (Fig. 6.34). By default, Simulink gives the name Scope, Scope1, Scope2, ... to the scope blocks.

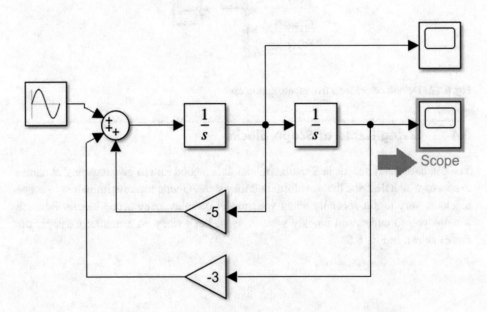

Fig. 6.34 Name of the lower scope block is "Scope"

Click on the appeared name shown in Fig. 6.34 and change it to x1 (Fig. 6.35).

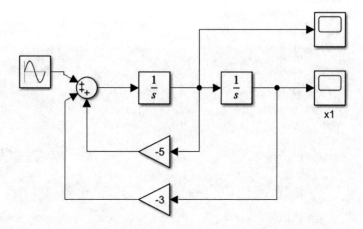

Fig. 6.35 Name of lower scope block is changed to x1

Run the simulation and double click the scope x1. As shown in Fig. 6.36, name of the scope block is appeared on the title bar. So, you can easily understand that this is signal x1 which you see.

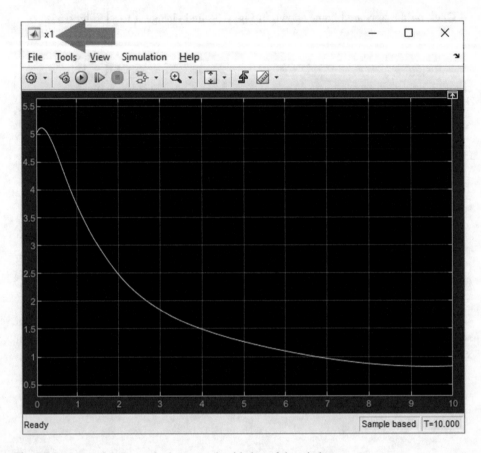

Fig. 6.36 Name of the scope is shown on the title bar of the window

6.5   Selection of Solver

Simulink used the ode45 to solve the differential equation that we studied in our first example. Simulink permits you to change the solver to what you like. Press Ctrl+E in order to change the solver. After pressing the Ctrl+E, the window shown in Fig. 6.37 appears on the screen. Click the solver details (Fig. 6.37). Now the window changes to what shown in Fig. 6.38.

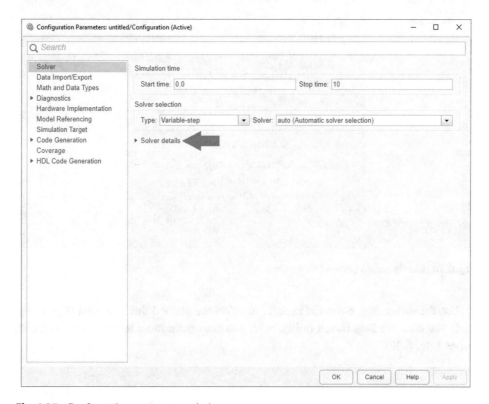

Fig. 6.37 Configuration parameters window

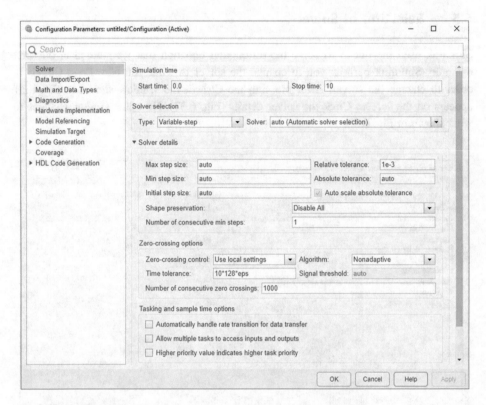

Fig. 6.38 Configuration parameters window

Use the solver drop down list in order to select the method that you want (Fig. 6.39).

If you click the help button in Fig. 6.39, you can obtain more information about each solver (Fig. 6.40).

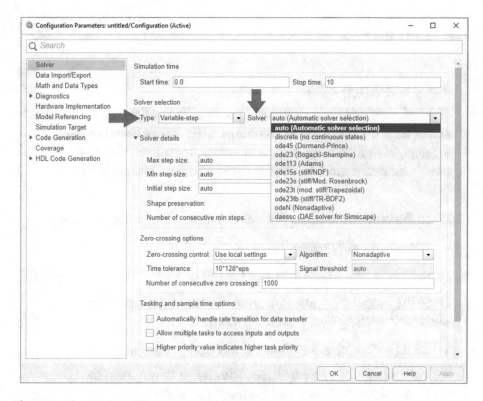

Fig. 6.39 Simulink has different types of solvers

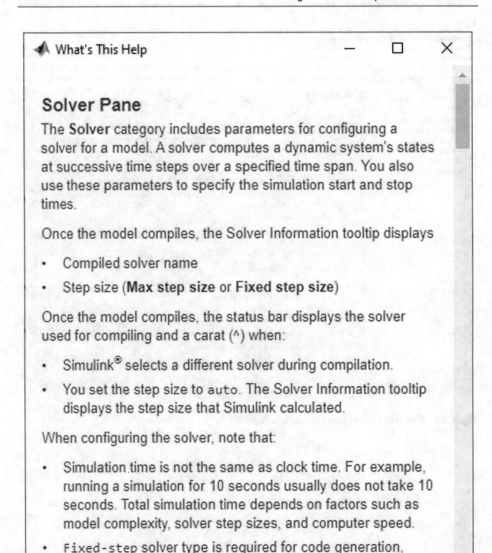

Fig. 6.40 Help page

6.6   Transferring the Results from Simulink Environment to MATLAB®

In the previous sections we drew the Simulink model and we observed the waveforms with the aid of scope block. The waveforms can be transferred to MATLAB environment and can be further processed there as well. In this section we learn how to transfer from Simulink to MATLAB environment.

Add a to workspace block (Fig. 6.41) to the Simulink model of first example of this chapter (Fig. 6.42).

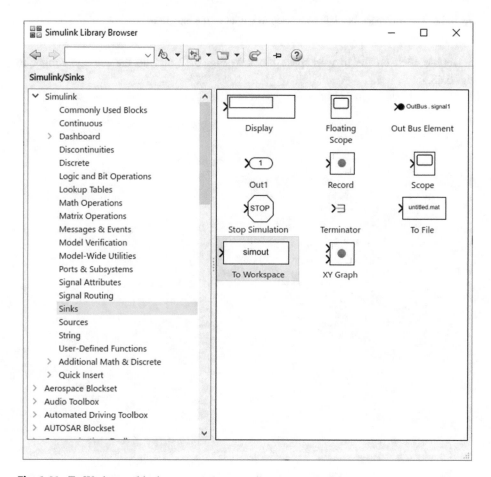

Fig. 6.41   To Workspace block

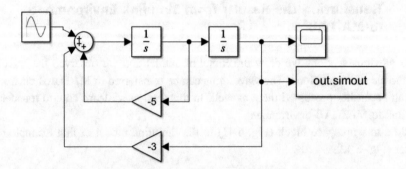

Fig. 6.42 To Workspace block is connected to output of the integrator

Double click on the to workspace block in Fig. 6.42 and do the settings similar to Fig. 6.43.

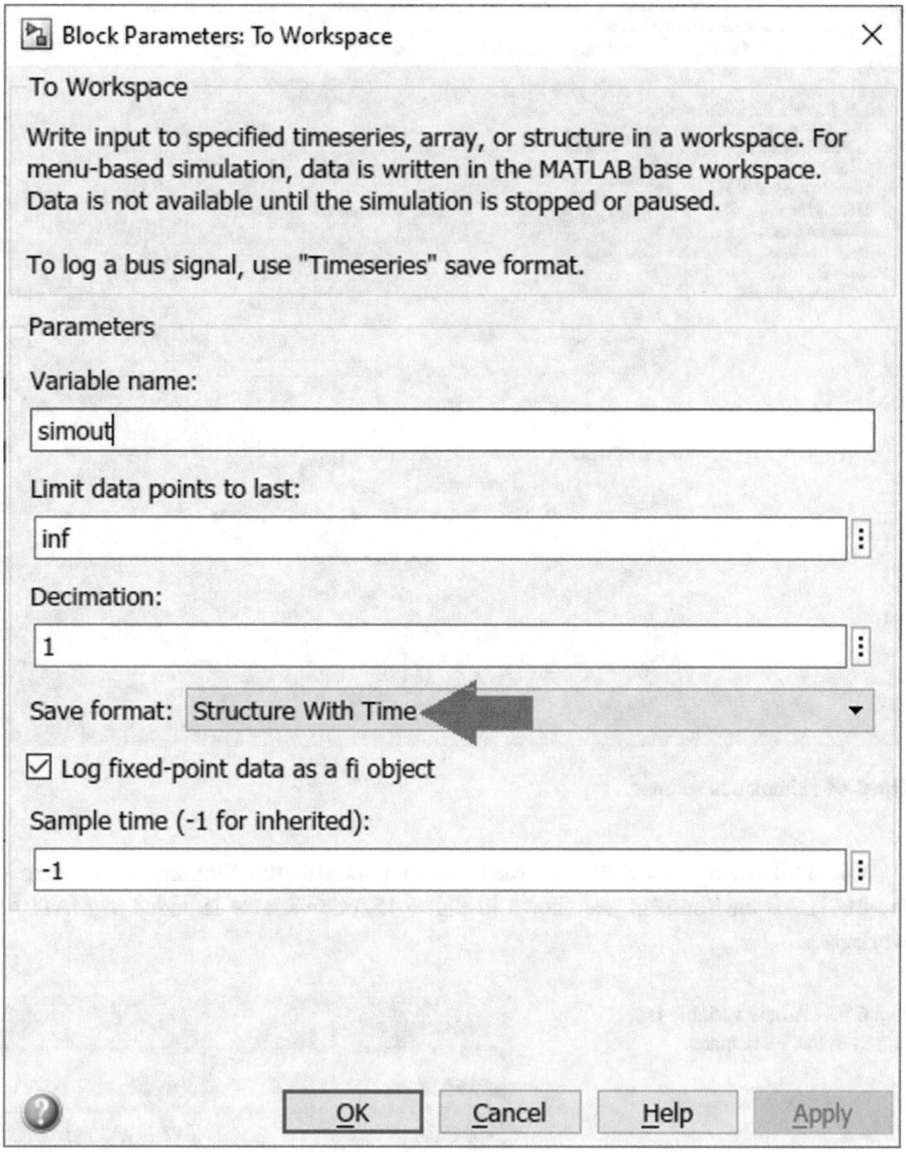

Fig. 6.43 Settings of to workspace block

Press the Ctrl+E and do the settings similar to Fig. 6.44. This asks the Simulink to solve the differential equation with fixed steps. Step size is $\Delta t = 0.01$ s. Click the OK button in Fig. 6.44. Then run the simulation.

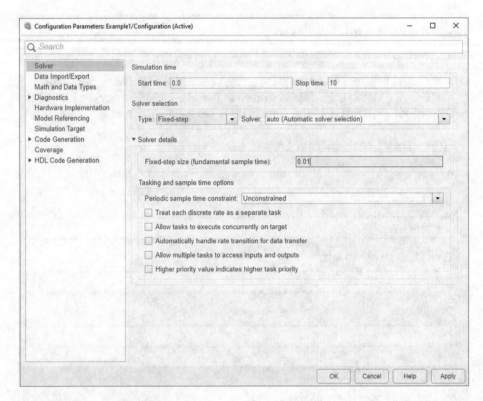

Fig. 6.44 Simulation settings

Wait until Simulink finishes its work. Then minimize the Simulink and return to the MATLAB environment. As shown in Fig. 6.45, one variable is added to MATLAB workspace.

Fig. 6.45 A new variable is added to the Workspace

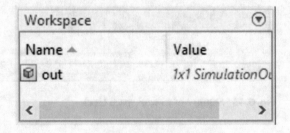

Use the commands shown in Fig. 6.46 in order to obtain the time and output values.

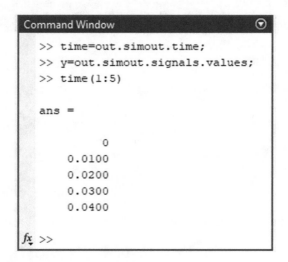

Fig. 6.46 Extraction of time and values of the signal from variable out

The commands shown in Fig. 6.47, shows the values of first 5 elements of vector time. Note that the time difference between two consecutive elements is 0.01 s as expected. Output values are transferred to variable y.

```
Command Window                          ⊙
    >> time=out.simout.time;
    >> y=out.simout.signals.values;
    >> time(1:5)

    ans =

               0
        0.0100
        0.0200
        0.0300
        0.0400

fx >>
```

Fig. 6.47 First five element of variable time

The commands shown in Fig. 6.48 draws the graph of imported data. Output of this code is shown in Fig. 6.49.

```
Command Window                                    ⊙
  >> time=out.simout.time;
  >> y=out.simout.signals.values;
  >> plot(time,y)
fx >>
```

Fig. 6.48 Plotting the imported data

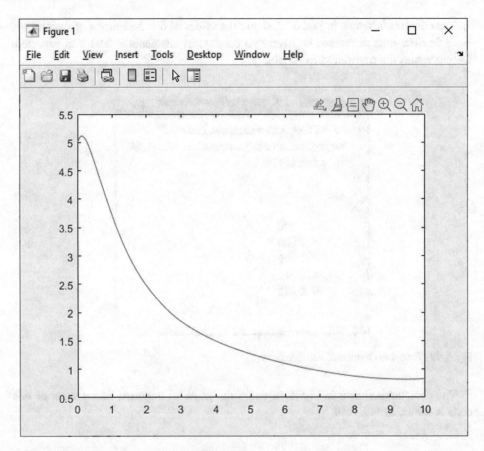

Fig. 6.49 Output of the code in Fig. 6.48

6.7 Transfer Function Block

In the first example of this chapter you learned how to convert a given differential equation into a state space model. When the given initial condition is zero, you can use the transfer function block to solve the differential equation and there is no need to convert the given equation into the state space equations. For instance, assume that we want to obtain the step response of the following system:

$$\ddot{y}(t) + 5\dot{y}(t) + 3y(t) = u(t),\ y(0) = 0,\ \dot{y}(0) = 0 \tag{6.6}$$

Let's start. Drag and drop a transfer fcn block (Fig. 6.50) to the working area and connect a scope block to its output (Fig. 6.51).

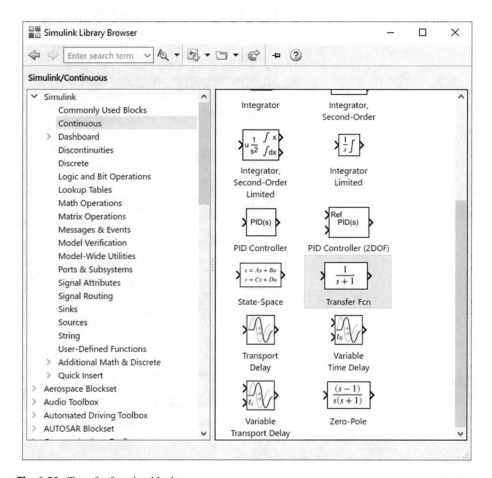

Fig. 6.50 Transfer function block

Fig. 6.51 An scope block is connected to the output of the transfer function block

In this example $\ddot{y}(t) + 5\dot{y}(t) + 3y(t) = u(t)$ so $Y(s)(s^2 + 5s + 3) = U(s)$ or $H(s) = \frac{Y(s)}{U(s)} = \frac{1}{s^2 + 5s + 3}$. Note that $H(s)$ shows the transfer function of the system. Double click the transfer function block in Fig. 6.51 and enter the coefficients of $H(s)$. After clicking the OK button the Simulink model changes to what shown in Fig. 6.53 (Fig. 6.52).

Fig. 6.52 Settings of transfer function block

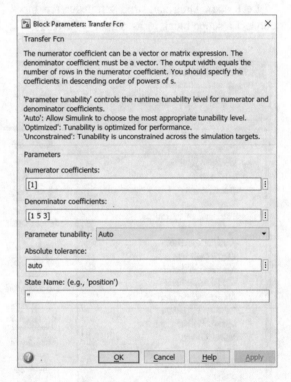

Fig. 6.53 Transfer function is changed to $\frac{1}{s^2 + 5s + 3}$

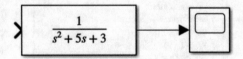

Drag and drop a step block (Fig. 6.54) to the Simulink model. Then connect the step block to the input of the transfer function (Fig. 6.55).

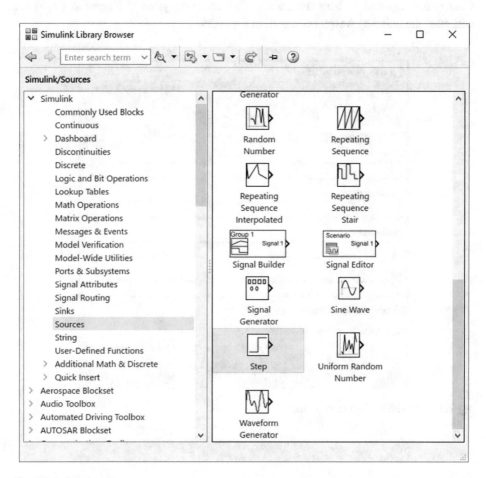

Fig. 6.54 Step block

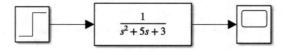

Fig. 6.55 Step block is connected to the input of transfer function block

Double click the step block in Fig. 6.55. Note that step time box is filled with 1 by default (Fig. 6.56). This cause the step block to generate the waveform shown in Fig. 6.57. Change the step time to zero (Fig. 6.58). This cause the jump to happen at t = 0 and generates the unit step waveform shown in Fig. 6.59.

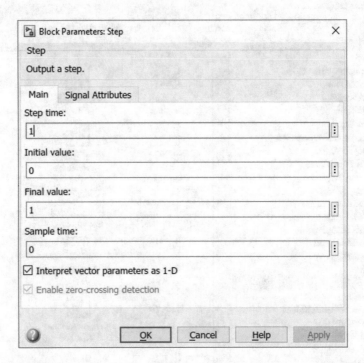

Fig. 6.56 Default settings of step block

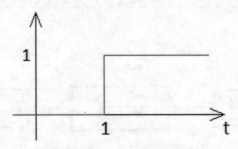

Fig. 6.57 Waveform generated with settings shown in Fig. 6.56

Fig. 6.58 Step time is changed to 0

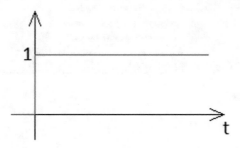

Fig. 6.59 Waveform generated with settings shown in Fig. 6.58

The simulation is run with the settings shown in Fig. 6.60 (press the Ctrl+E in order open the configuration parameters window). After running the simulation, double click the scope block to open it. Waveform of the scope block is shown in Fig. 6.61.

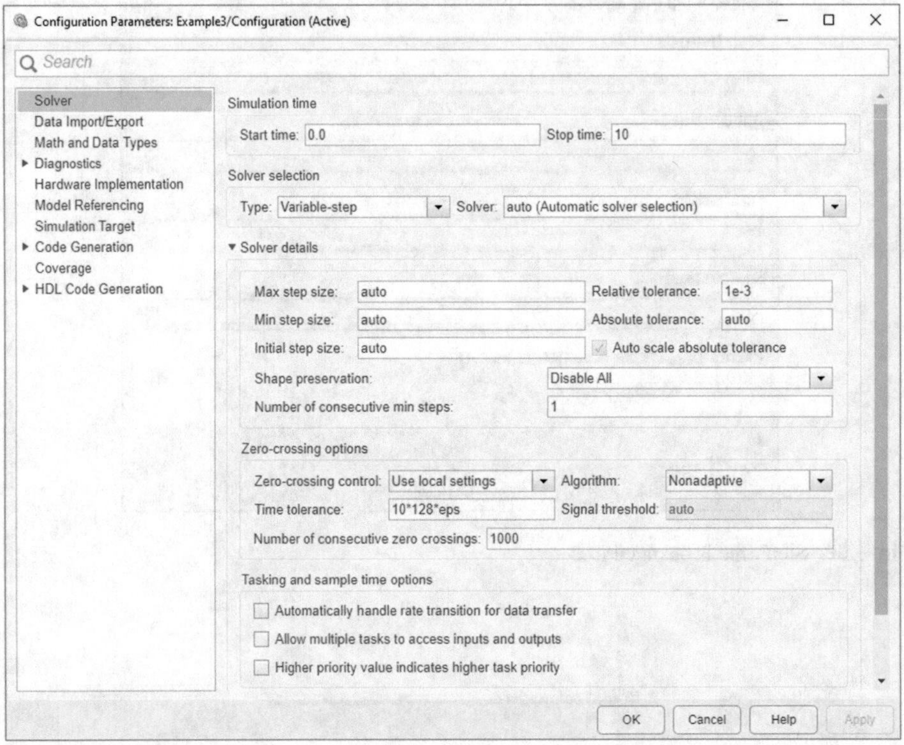

Fig. 6.60 Solver settings

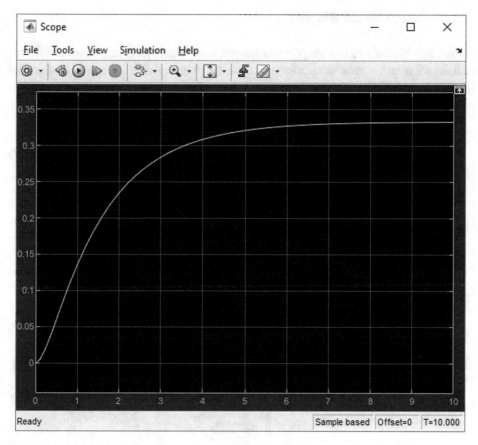

Fig. 6.61 Output of simulation

6.8   State Space Block

In the first example of this chapter you learned how to convert a time domain differential equation into a state space equation. State space equations are required to draw the Simulink model. Now, assume that you are given a linear state space model directly. For instance, consider the following system:

$$\begin{cases} \dot{x}_1 = 0x_1 + 1x_2 + 1u \\ \dot{x}_2 = -2x_1 - 5x_2 + 2u \end{cases}, \quad \begin{bmatrix} x_1(0) \\ x_2(0) \end{bmatrix} = \begin{bmatrix} 2 \\ 3 \end{bmatrix} \tag{6.7}$$
$$y = 1x_1 + 3x_2 + 0u$$

In the above equation x_1 and x_2 show the state variables, u shows the input and y shows the output. In this example $u = 0$. The above system can be written as $\begin{cases} \dot{x} = Ax + Bu \\ y = Cx + Du \end{cases}$

where $x = \begin{bmatrix} x_1 \\ x_2 \end{bmatrix}$, $A = \begin{bmatrix} 0 & 1 \\ -2 & -5 \end{bmatrix}$, $B = \begin{bmatrix} 1 \\ 2 \end{bmatrix}$, $C = \begin{bmatrix} 1 & 3 \end{bmatrix}$ and $D = [0]$. Such a linear system can be simulated with the aid of state-space block (Fig. 6.62).

Fig. 6.62 State-space block

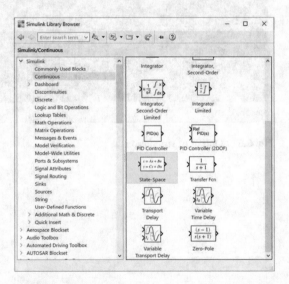

Draw the Simulink model shown in Fig. 6.63. Input of the state-space block is connected to a constant block (Fig. 6.64) with value of zero. Settings of the constant block is shown in Fig. 6.65.

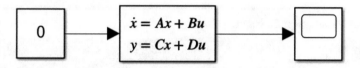

Fig. 6.63 Simulink diagram to simulate the given equation

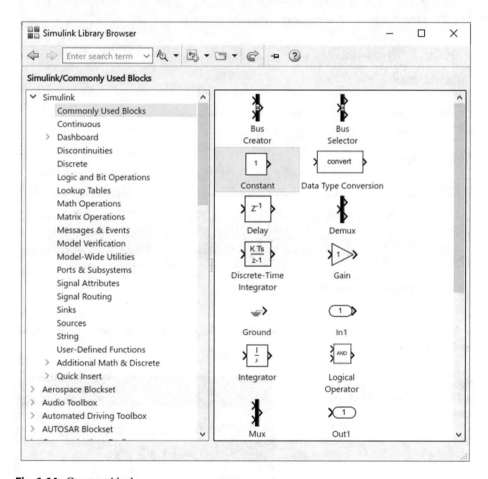

Fig. 6.64 Constant block

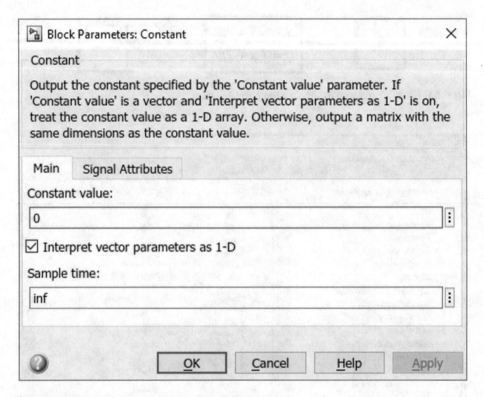

Fig. 6.65 Settings of constant block in Fig. 6.63

Minimize the Simulink and define the A, B, C and D variables in MATLAB environment (Fig. 6.66).

Fig. 6.66 Variables A, B, C and D are defined in MATLAB work space

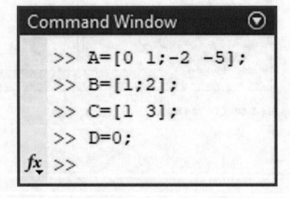

Return to the Simulink environment and double click the state-space block. Do the settings similar to Fig. 6.67. Note that Simulink can use variables that are defined in MAT-LAB environment. You can enter the values of A, B, C and D directly to the textboxes as well (Fig. 6.68).

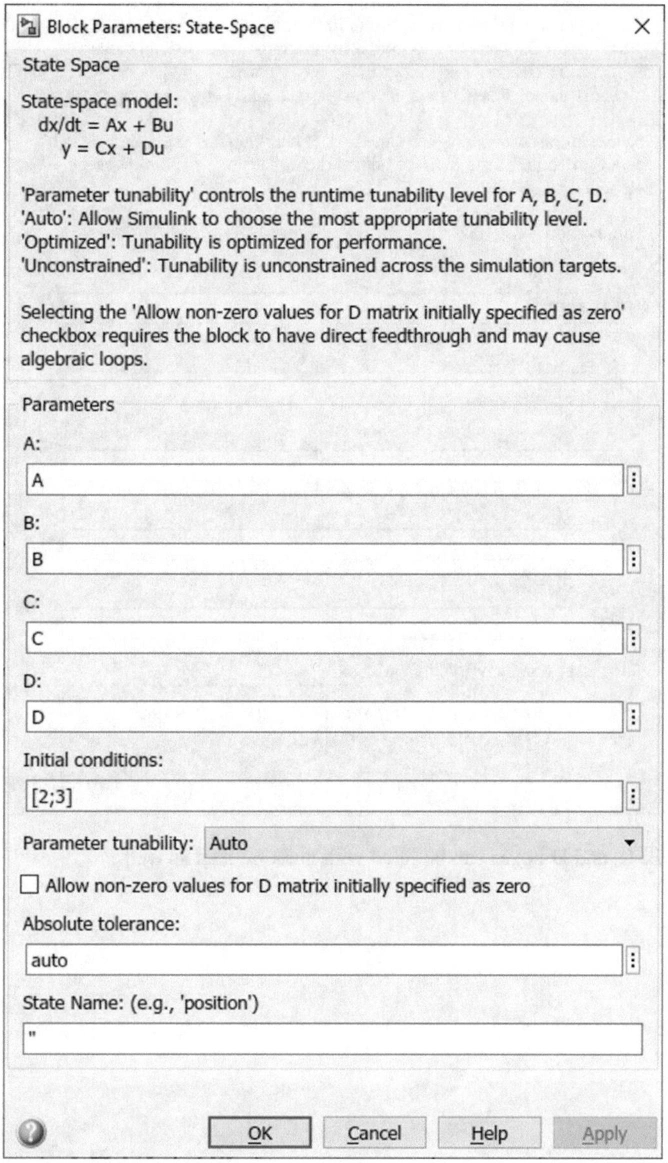

Fig. 6.67 Settings of state space block

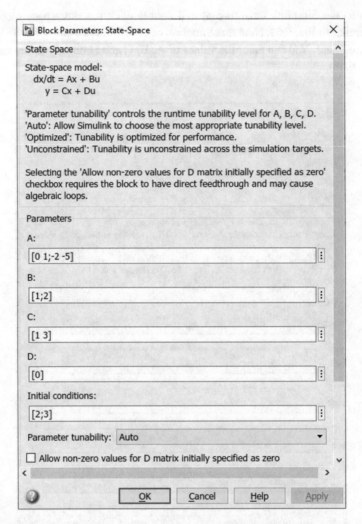

Fig. 6.68 A, B, C, and D boxes can be filled with numeric data as well

Now run the simulation. Simulation result is shown in Fig. 6.69. The waveform which is shown in Fig. 6.69 is the output of the system, $y(t)$.

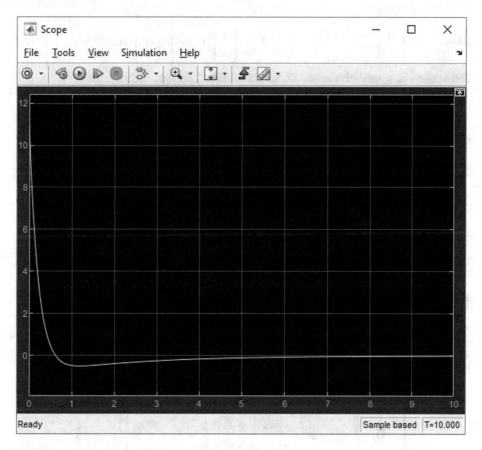

Fig. 6.69 Output of simulation

Assume that we want to observe the $x_1(t)$ and $x_2(t)$ waveform as well. In this case we need to add some rows to the output and update it:

$$\begin{cases} \dot{x}_1 = 0x_1 + 1x_2 + 1u \\ \dot{x}_2 = -2x_1 - 5x_2 + 2u \end{cases}, \quad \begin{bmatrix} x_1(0) \\ x_2(0) \end{bmatrix} = \begin{bmatrix} 2 \\ 3 \end{bmatrix}$$

$$y = \begin{bmatrix} 1x_1 + 3x_2 \\ x_1 \\ x_2 \end{bmatrix} + \begin{bmatrix} 0 \\ 0 \\ 0 \end{bmatrix} u \qquad (6.8)$$

In the above equation x_1 and x_2 show the state variables, u shows the input and y shows the output. As told $u = 0$. The above system can be written as $\begin{cases} \dot{x} = Ax + Bu \\ y = Cx + Du \end{cases}$

where $x = \begin{bmatrix} x_1 \\ x_2 \end{bmatrix}$, $A = \begin{bmatrix} 0 & 1 \\ -2 & -5 \end{bmatrix}$, $B = \begin{bmatrix} 1 \\ 2 \end{bmatrix}$, $C = \begin{bmatrix} 1 & 3 \\ 1 & 0 \\ 0 & 1 \end{bmatrix}$ and $D = \begin{bmatrix} 0 \\ 0 \\ 0 \end{bmatrix}$. Let's

define these new variables in MATLAB environment (Fig. 6.70).

```
Command Window                        ⊙

>> A=[0 1;-2 -5];
>> B=[1;2];
>> C=[1 3;1 0;0 1];
>> D=[0;0;0];
fx >>
```

Fig. 6.70 New values for matrix C and D are entered

Now run the model (Fig. 6.71). Output of the simulation is shown in Fig. 6.72. As you see, three waveforms are appeared on the scope screen. The yellow one is $1x_1 + 3x_2 + 0u$, the blue one is x_1 and the red one is x_2.

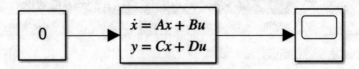

Fig. 6.71 Simulink diagram to simulate the given model

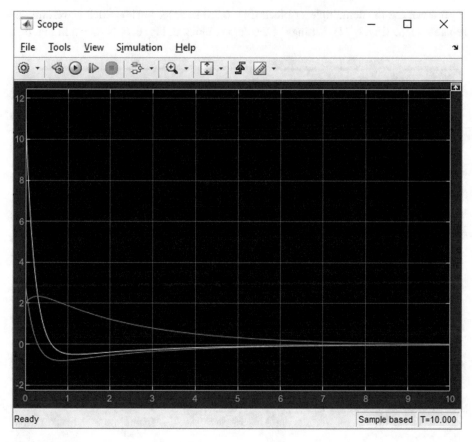

Fig. 6.72 Output of simulation

You can use the demultiplexer block (Fig. 6.73) in order to show each waveform on a separate scope (Fig. 6.74). Settings of the demux block in Fig. 6.74 is shown in Fig. 6.75.

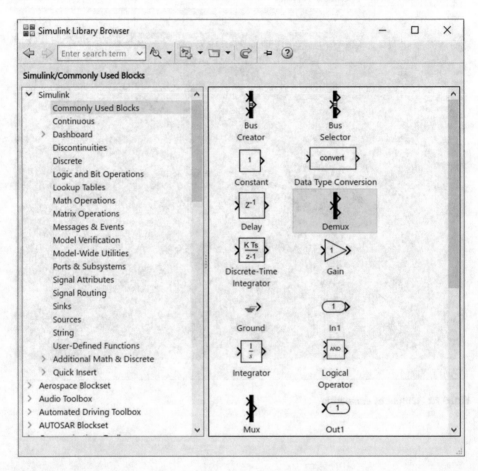

Fig. 6.73 Demultiplexer block

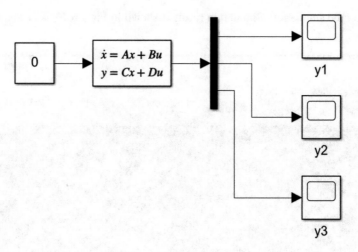

Fig. 6.74 Demultiplexer block is connected to the Simulink diagram

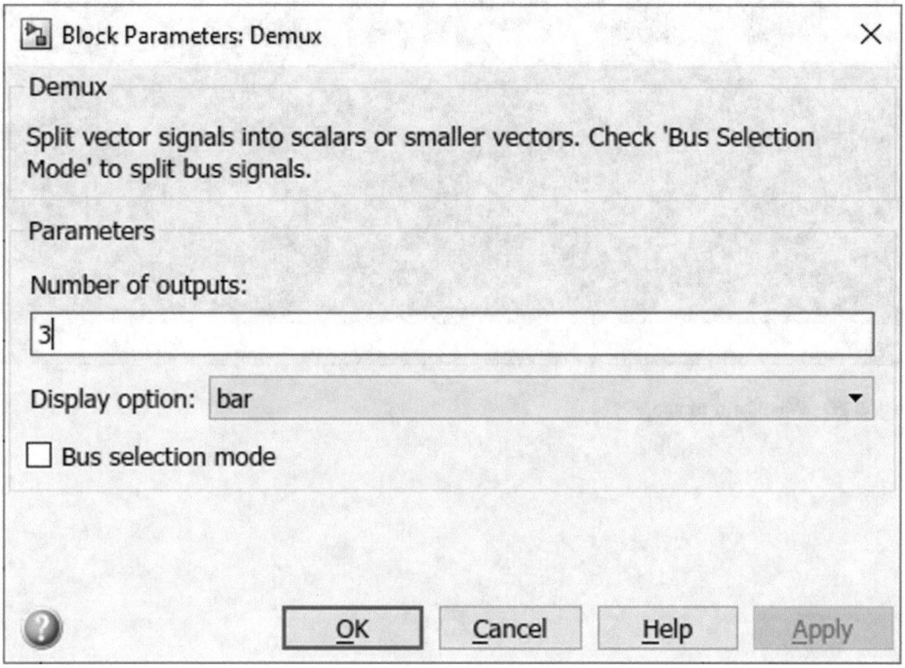

Block Parameters: Demux   ✕

Demux

Split vector signals into scalars or smaller vectors. Check 'Bus Selection Mode' to split bus signals.

Parameters

Number of outputs:

3

Display option: bar ▼

☐ Bus selection mode

OK   Cancel   Help   Apply

Fig. 6.75 Settings of demultiplexer block

Now run the simulation. Simulation result is shown in Figs. 6.76, 6.77 and 6.78.

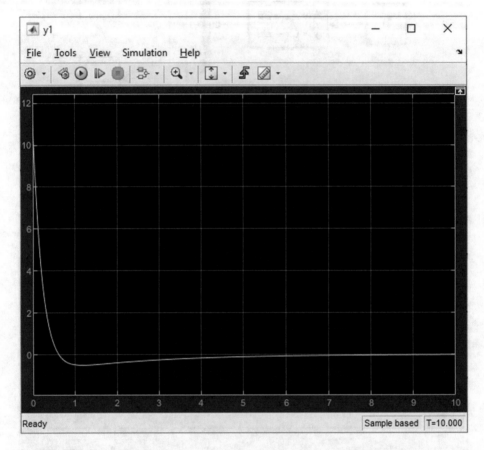

Fig. 6.76 Waveform of scope y1

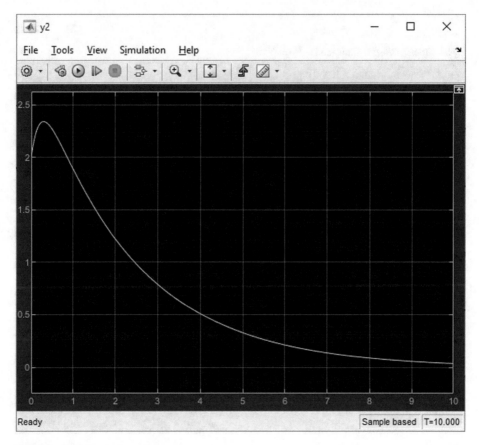

Fig. 6.77 Waveform of scope y2

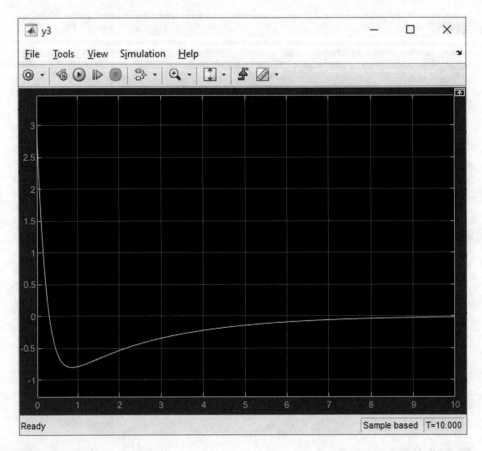

Fig. 6.78 Waveform of scope y3

6.9   Nonlinear Models

In the previous examples we learned how to prepare the Simulink model of the linear systems. In this section we will learn how to prepare the Simulink model for non-linear systems. Consider the following non-linear equation:

$$\ddot{y}(t) + 5\dot{y}(t) + 3y(t)^2 = u(t), \quad y(0) = 5, \quad \dot{y}(0) = 2$$
$$u(t) = 3 + 0.7 \, \sin\left(0.5t + \tfrac{\pi}{4}\right)$$

(6.9)

Let's convert the model into the state-space:

$$\begin{cases} x_1(t) = y(t) \\ x_2(t) = \dot{y}(t) \end{cases} \Rightarrow \begin{cases} \dot{x}_1(t) = x_2(t) \\ \dot{x}_2(t) = -3x_1(t)^2 - 5x_2(t) + u(t) \end{cases}$$

(6.10)

Implementation of obtained state space model is shown in Fig. 6.79. Multiplication is implemented with a divide block. Settings of the divide block (Fig. 6.80) is shown in Fig. 6.81. Output of the divide block generates the $x_1(t)^2$ term for us.

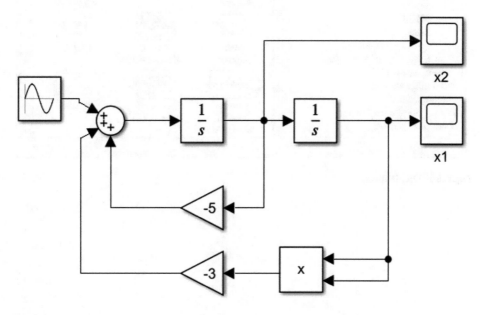

Fig. 6.79 Simulink model of given nonlinear equation

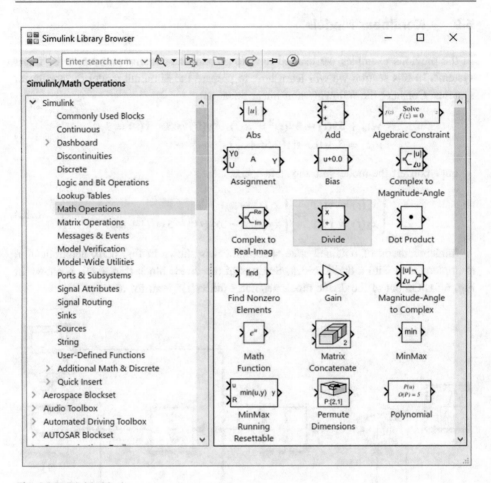

Fig. 6.80 Divide block

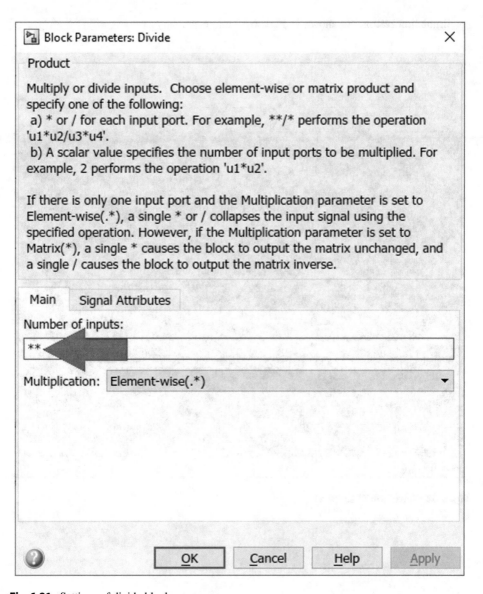

Fig. 6.81 Settings of divide block

Simulation results are shown in Figs. 6.82 and 6.83.

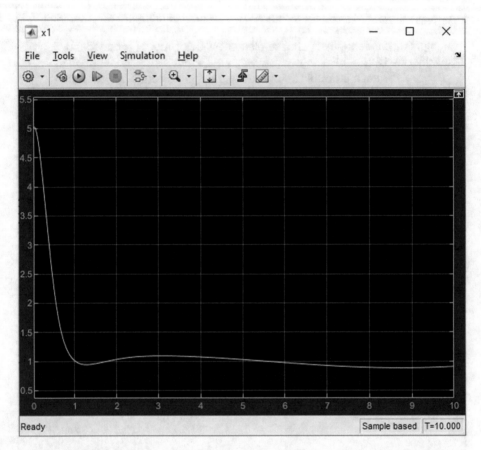

Fig. 6.82 Waveform of scope x1

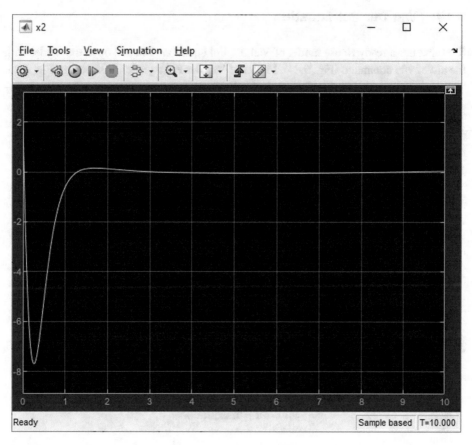

Fig. 6.83   Waveform of scope x2

6.10   Van Der Pol Equation

Simulink has a ready to use model of Van der Pol equation. You can open the model with the aid of vdp command (Fig. 6.84). Model of Van der Pol equation is shown in Fig. 6.85. This model implements the $\frac{d^2y}{dt^2} + \mu(1 - y^2)\frac{dy}{dt} + t = 0$ equation which is a nonlinear equation. The Mu gain block determines the value of μ. You can double click on the integrators and enter the desired initial condition. For instance, output of the system for $\mu = 0.2$ and $\begin{bmatrix} x_1(0) \\ x_2(0) \end{bmatrix} = \begin{bmatrix} 2 \\ 0 \end{bmatrix}$ is shown in Fig. 6.86.

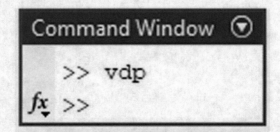

Fig. 6.84 vdp command

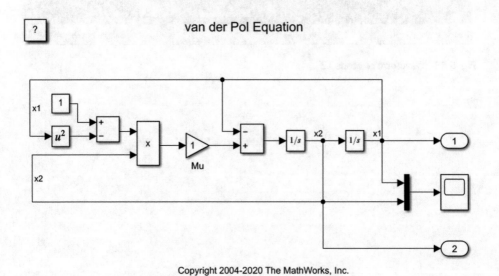

Fig. 6.85 Simulink model of Van der Pol equation

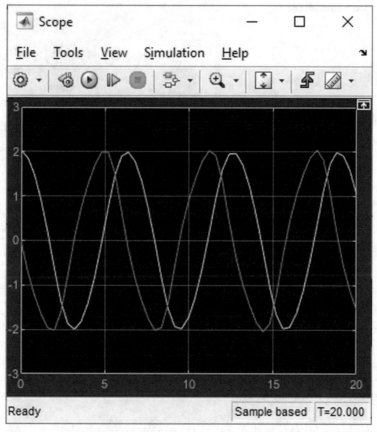

Fig. 6.86 Waveform for $\mu = 0.2$ and $\begin{bmatrix} x_1(0) \\ x_2(0) \end{bmatrix} = \begin{bmatrix} 2 \\ 0 \end{bmatrix}$

Exercises

1. Draw the Simulink diagram of $\frac{d^2y(t)}{dt} + 6\frac{dy(t)}{dt} + y(t) = \frac{du(t)}{dt} + 6u(t)$. Note that $y(t)$ and $u(t)$ show the system output and input, respectively. **Hint**: You can use the derivative block (Fig. 6.87).
2. Assume that $u(t) = \sin(t)$. Use Simulink to observe the output of the system given in exercise 1. All the initial conditions are assumed to be zero.
3. Use the transfer function block (Fig. 6.50) in order to obtain the step response of the system given in exercise 1.

Fig. 6.87 Derivative block

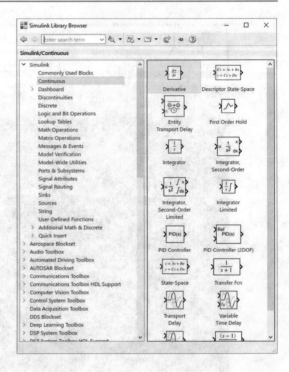

References for Further Study

1. Beginning MATLAB and Simulink: From Novice to Professional, Eshkabilov S., Apress, 2019.
2. Practical MATLAB Modelling with Simulink: Programming and Simulating Ordinary and Partial Differential Equations, Eshkabilov S., Apress, 2020.

Solving Difference Equations in Simulink® 7

7.1 Introduction

Differential equations deal with continuous system, while the difference equations are meant for discrete process. Generally, a difference equation is obtained in an attempt to solve an ordinary differential equation by finite difference method. Thus, a difference equation is a relation between the differences of unknown function at one or more general values of the independent variable. In this chapter you will learn how to solve difference equations in the Simulink environment.

7.2 Solving the Difference Equations: Example 1

In this example we want to solve the following difference equation with Simulink

$$2 \times y(n-1) + 6 \times y(n-2) = y(n), \quad y(-1) = 11 \quad y(-2) = 7 \tag{7.1}$$

Let's calculate first few terms:

$$y(0) = 2 \times 11 + 6 \times 7 = 64$$
$$y(1) = 2 \times 64 + 6 \times 11 = 194$$
$$y(2) = 2 \times 194 + 6 \times 64 = 772 \tag{7.2}$$

The building block to make discrete models is the delay block (Fig. 7.1). Effect of this block on the signal is shown in Fig. 7.2.

© The Author(s), under exclusive license to Springer Nature Switzerland AG 2023
F. Asadi, *Applied Numerical Analysis with MATLAB®/Simulink®*,
Synthesis Lectures on Engineering, Science, and Technology,
https://doi.org/10.1007/978-3-031-19366-8_7

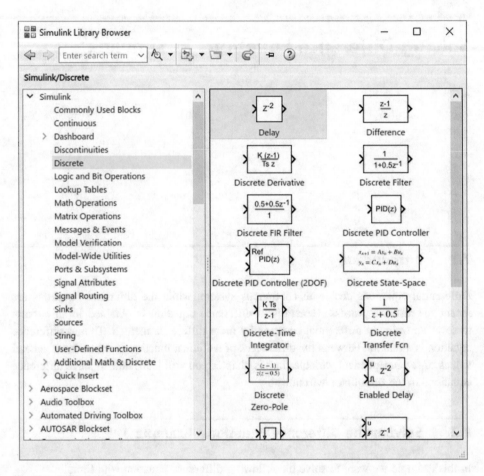

Fig. 7.1 Delay block

Fig. 7.2 Relation between
input and output of a delay
block

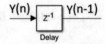

Draw the block diagram shown in Fig. 7.3. Figure 7.4 is the same as Fig. 7.3, however some labels are added to it in order to easily see that this block diagram implements the $2 \times x(n-1) + 6 \times x(n-2) = x(n)$ equation.

Fig. 7.3 Simulink model for
$2 \times y(n-1) + 6 \times y(n-2) =$
$y(n)$

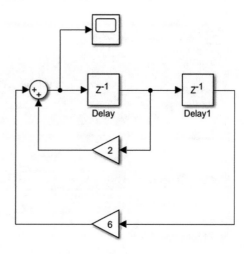

Fig. 7.4 Labels are added to
the Simulink model shown in
Fig. 7.3

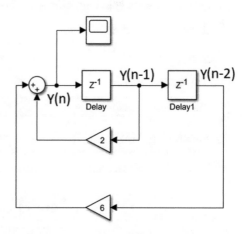

It's time to enter the initial conditions. Settings of delay blocks Delay and Delay1 are
shown in Figs. 7.5 and 7.6, respectively. Delay length box determines the length of delay.
For instance, if you want to generate $x(n-5)$ from $x(n)$, then you need to set the Delay
length box equal to 5.

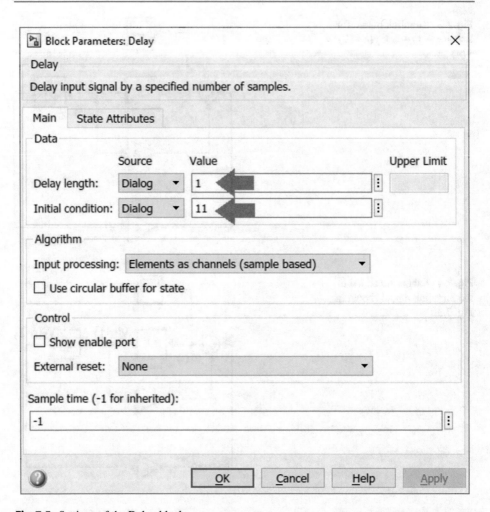

Fig. 7.5 Settings of the Delay block

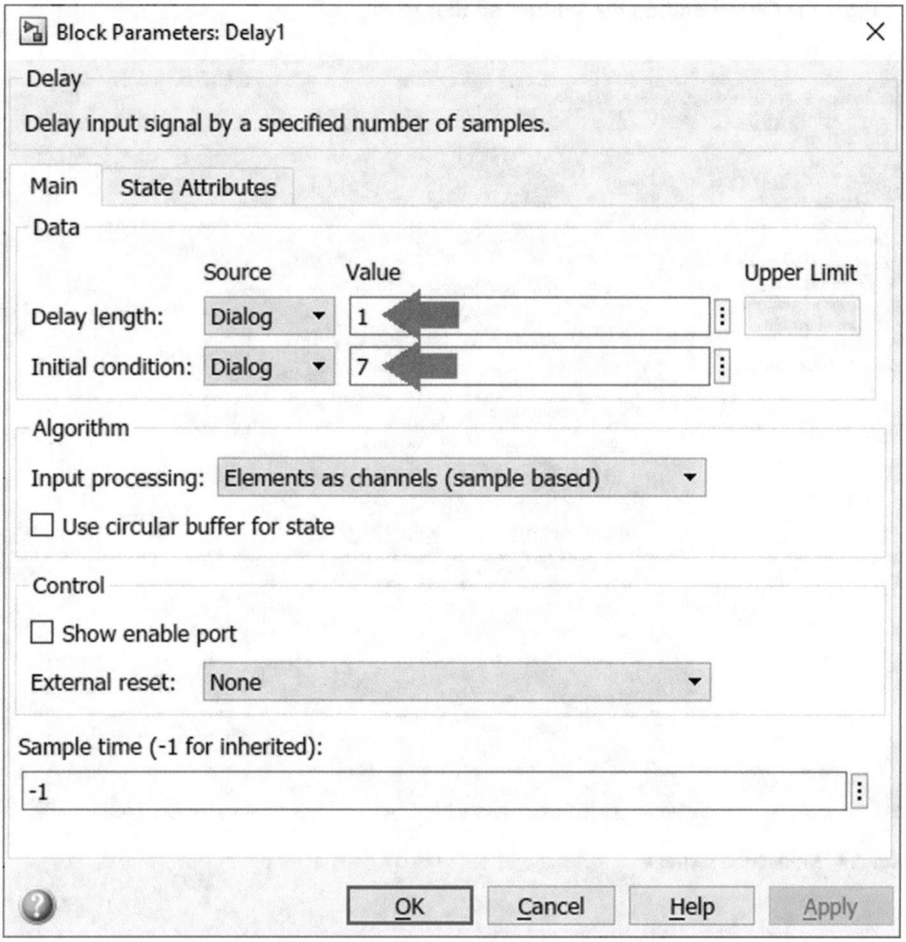

Fig. 7.6 Settings of the Delay1 block

Press the Ctrl+E and do the settings similar to Fig. 7.7.

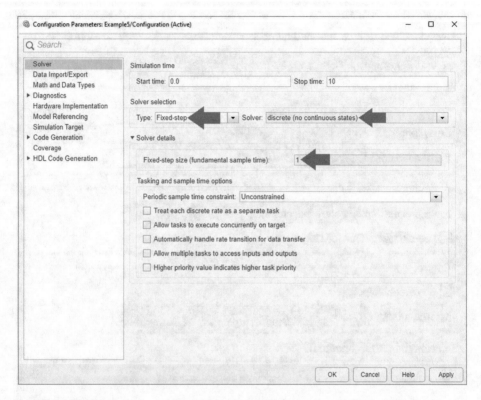

Fig. 7.7 Simulation settings

Enter the desired stop time and run the simulation. In this example we want to calculate $x(0)$, $x(1)$ and $x(2)$. So, stop time is filled with 3 (Fig. 7.8). Simulation result is shown in Fig. 7.9. Compare the result with values we found using hand calculation.

Fig. 7.8 Stop time is entered

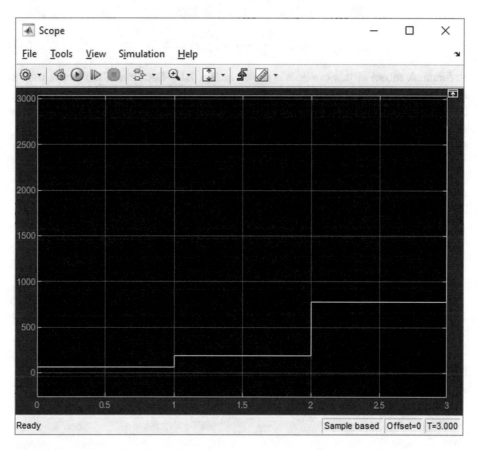

Fig. 7.9 Simulation result

7.3   Solving the Difference Equations: Example 2

In this example we want to solve the following equation:

$$y(n) - 3 \times y(n-1) = 4^{n-1}, y(-1) = 0 \tag{7.3}$$

We can rewrite the given equation as:

$$y(n) = 3 \times y(n-1) + 4^{n-1}, y(-1) = 0 \tag{7.4}$$

Let's calculate the first few terms:

$$y(0) = 3 \times 0 + 4^{-1} = 0.25,$$
$$y(1) = \frac{3}{4} + 1 = \frac{7}{4} = 1.75,$$
$$y(2) = \frac{21}{4} + 4 = \frac{37}{4} = 9.25 \tag{7.5}$$

Simulink model of this example is shown in Fig. 7.10. Location of the blocks used in this model are shown in Figs. 7.11, 7.12 and 7.13. Zero-Order Hold block is quite similar to the bracket function. Remember that bracket function removes the decimal part of the number, for instance, $[1.2] = 1$ and $[-1.2] = -2$.

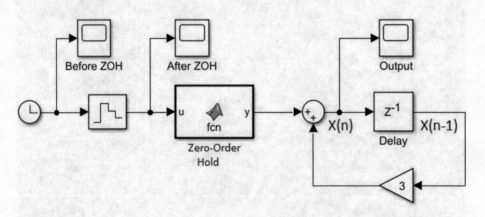

Fig. 7.10 Simulink model for second example

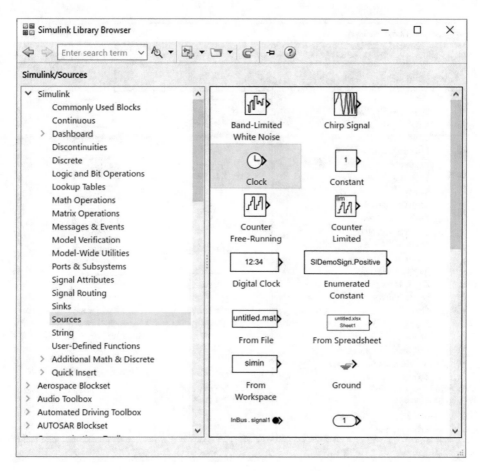

Fig. 7.11 Clock block

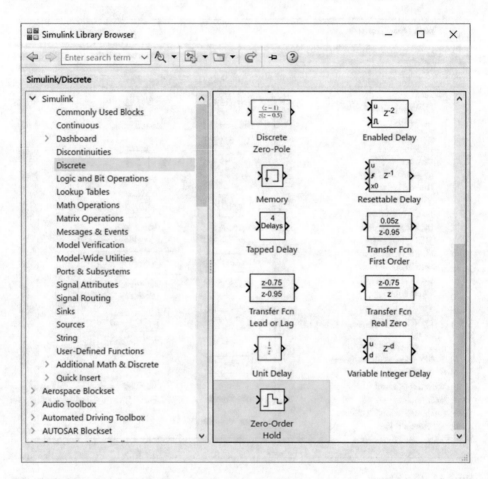

Fig. 7.12 Zero order hold block

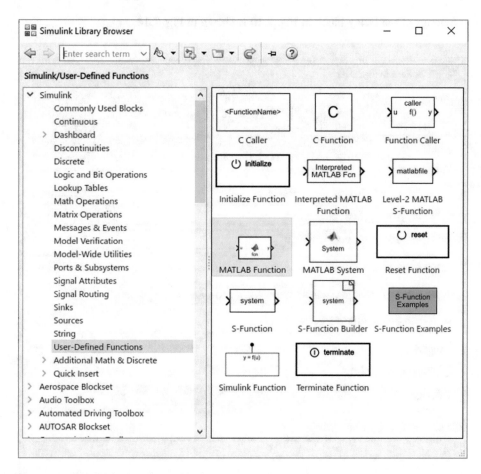

Fig. 7.13 MATLAB function block

Settings of the Delay block in Fig. 7.10 is shown in Fig. 7.14.

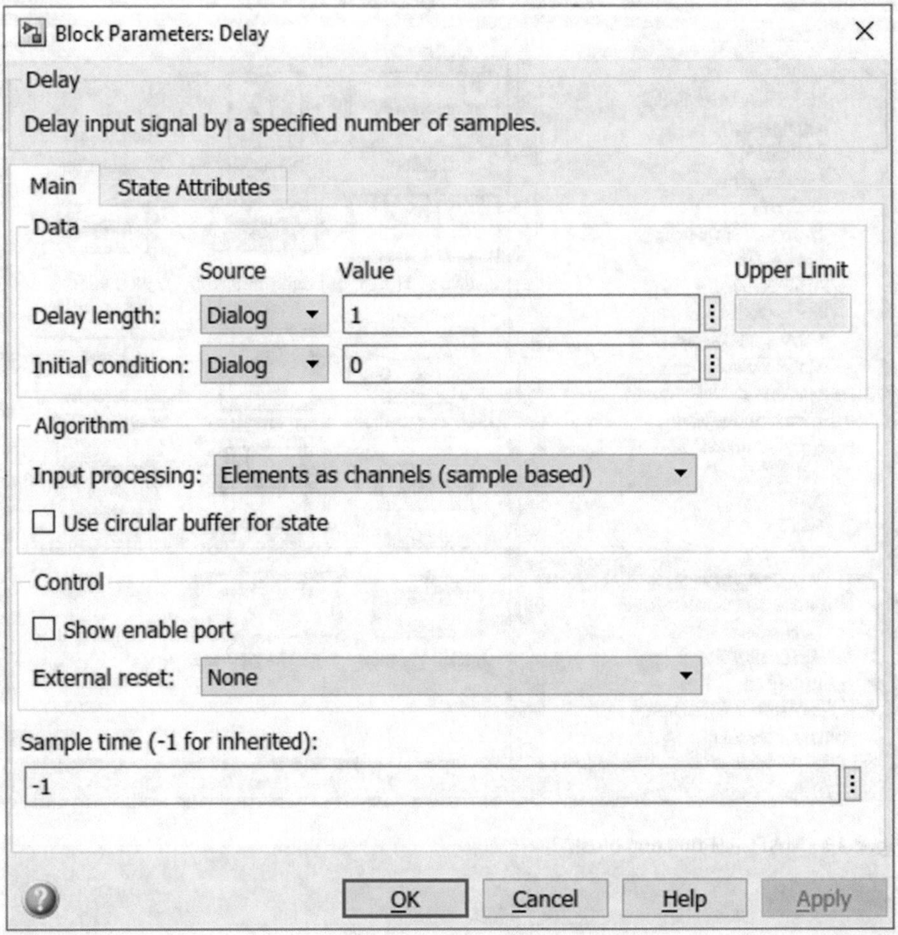

Fig. 7.14 Settings of Delay block

Double click the MATLAB function and enter the code shown in Fig. 7.15.

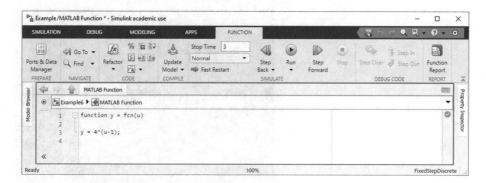

Fig. 7.15 Code for MATLAB function block

Run the simulation with the settings shown in Fig. 7.7. Simulation results are shown in Figs. 7.16, 7.17 and 7.18.

Fig. 7.16 Waveform of output scope

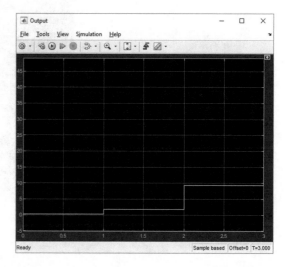

Fig. 7.17 Waveform of before
ZOH scope

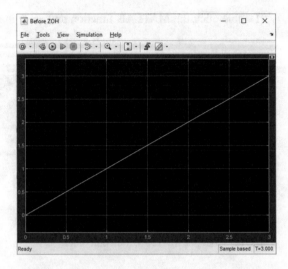

Fig. 7.18 Waveform of after
ZOH scope

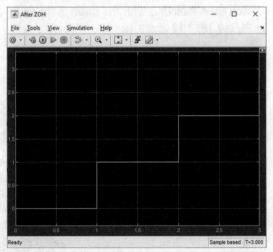

7.4   Solving the Difference Equations: Example 3

In this example we want to solve the following system:

$$y(n) - \frac{1}{2}y(n-1) = \frac{1}{4}x(n) + \frac{1}{4}x(n-1), \quad y(-1) = 0 \tag{7.6}$$

$y(n)$ and $x(n)$ shows the output and input of the system, respectively. $x(n)$ is given as:

$$x(n) = \begin{cases} 0 & n < 0 \\ 1 & n \geq 0 \end{cases} \tag{7.7}$$

Let's rewrite the equation as follows:

$$y(n) = \frac{1}{2}y(n-1) + \frac{1}{4}x(n) + \frac{1}{4}x(n-1), \quad y(-1) = 0 \tag{7.8}$$

Let's calculate the first few terms:

$$y(0) = \frac{1}{4} = 0.25$$

$$y(1) = \frac{5}{8} = 0.625$$

$$y(2) = \frac{13}{16} = 0.812$$

$$y(3) = \frac{29}{32} = 0.906 \tag{7.9}$$

Simulink model of the given equation is shown in Fig. 7.19.

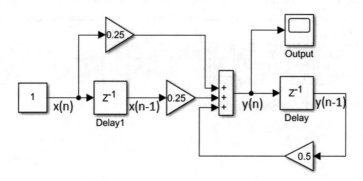

Fig. 7.19 Simulink model for third example

Settings for Delay1 and Delay blocks are shown in Fig. 7.20. Both of the blocks have settings similar to what shown in Fig. 7.20.

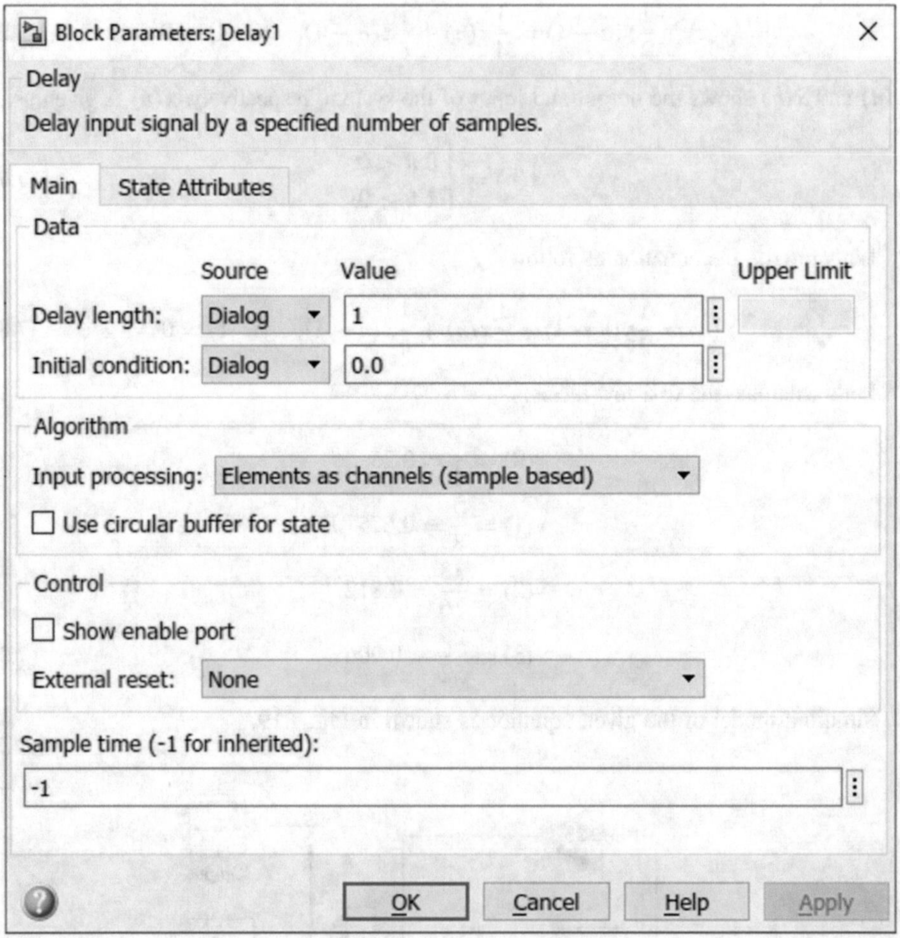

Fig. 7.20 Settings of Delay and Delay1 blocks

Settings of constant block is shown in Fig. 7.21. Sample time of the block is set to 1 s. So, the output of the block is samples each 1 s. The function which is sampled each 1 s is the constant function $f(t) = 1$ (remember that the block is constant). So, sampling makes an array of ones, i.e. 1, 1, 1, ... which simulates the role of $x(n) = \begin{cases} 0 & n < 0 \\ 1 & n \geq 0 \end{cases}$.

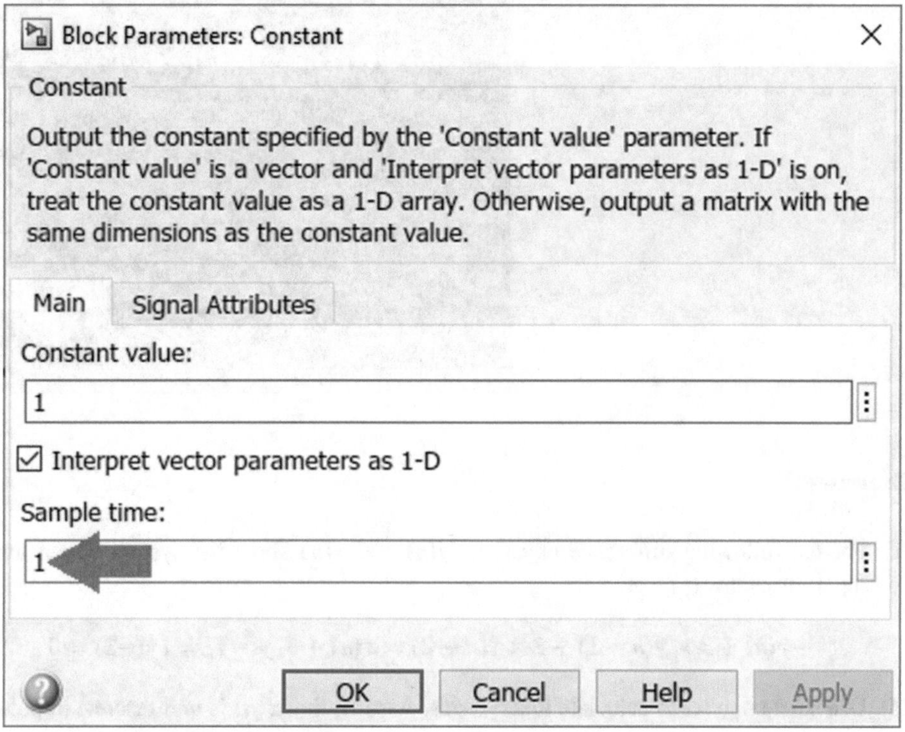

Fig. 7.21 Settings of Constant block

Run the simulation with stop time equal to 4 s and settings shown in Fig. 7.7. Simulation result is shown in Fig. 7.22. Compare it with the result of hand calculations done in the beginning of this example.

Fig. 7.22 Waveform of output scope

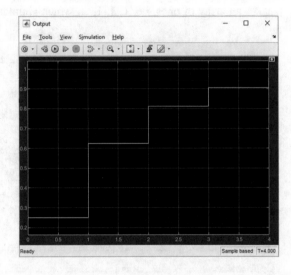

Exercises

1. For the following difference equation ($y(n)$ and $x(n)$ show the system output and input, respectively)

$$-y(n) + 3 \times y(n-1) + 2 \times y(n-2) = x(n) + 3, y(-1) = 1 \, y(-2) = 0$$

(a) Use hand analysis to calculate first 5 terms. Assume that $x(n)$ is unit discrete impulse function, i.e. $x(n) = \begin{cases} 1 \; n = 0 \\ 0 \; n \neq 0 \end{cases}$. **Hint**: You can use repeating sequence stair block (Fig. 7.23) to generate the given input. Use settings shown in Fig. 7.24.

(b) Use Simulink to solve the given equation and compare it with part (a).

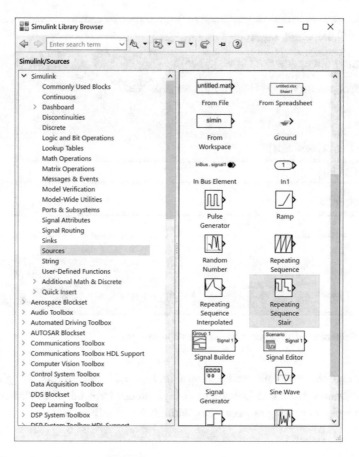

Fig. 7.23 Repeating sequence stair block

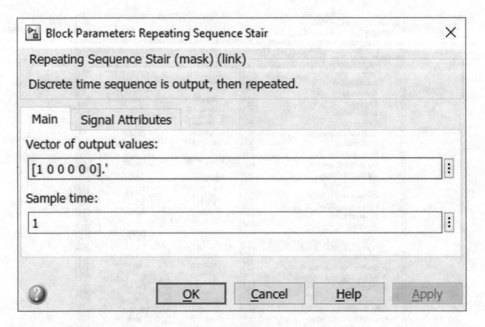

Fig. 7.24 Require settings to generate discrete impulse signal

(c) Show that the Simulink model shown in Fig. 7.25 implements the given differ-
    ence equation. Note that Delay and Delay1 blocks have initial conditions 1 and 0,
    respectively. Repeating sequence has settings shown in Fig. 7.24.

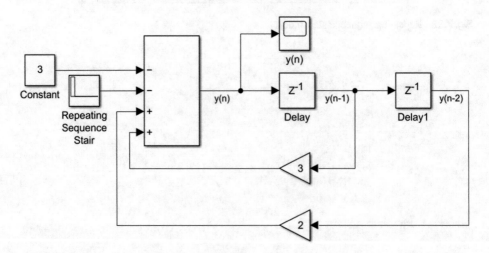

Fig. 7.25 Simulink model for part (c) of exercise 1

References for Further Study

1. Discrete-Time Signal Processing, Oppenheim AV., Schafer RW, Buck JR., Prentice Hall, 1998.
2. Discrete-Time Control Systems, Ogata K., Pearson, 1994.

Curve Fitting with MATLAB®

8

8.1 Introduction

Curve fitting is the process of constructing a curve, or mathematical function, that has
the best fit to a series of data points. In this chapter you will learn how to solve a curve
fitting problem with the aid of Curve Fitting Toolbox™ of MATLAB.

8.2 Example 1: Linear Curve Fitting

Consider the simple circuit shown in Fig. 8.1. This voltmeter and ammeter measures the
voltage and current of the resistor. Measured voltages and currents are shown in Table
8.1.

Fig. 8.1 Simple circuit to
measure voltage and current of
a resistor

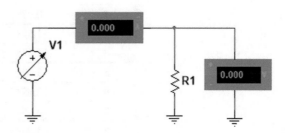

© The Author(s), under exclusive license to Springer Nature Switzerland AG 2023   191
F. Asadi, *Applied Numerical Analysis with MATLAB®/Simulink®*,
Synthesis Lectures on Engineering, Science, and Technology,
https://doi.org/10.1007/978-3-031-19366-8_8

Table 8.1 I–V data for circuit shown in Fig. 8.1

No.	V (V)	I (A)
1	0.579	0.10
2	0.978	0.17
3	1.598	0.28
4	1.976	0.34
5	2.496	0.43
6	2.953	0.51
7	3.458	0.60
8	4.068	0.71
9	4.450	0.78
10	4.917	0.86
11	5.35	0.93
12	5.75	1.01
13	6.37	1.11
14	6.60	1.15

We want to use Table 8.1 in order to calculate the resistance of R1. We can use the Ohm's law, i.e. $R = \frac{V}{I}$. Table 8.2 shows the measured resistance for each set of measured data. You can use the commands shown in Fig. 8.2 to obtain the values shown in Table 8.2.

Table 8.2 Calculation of V/I for each measured pair

No.	V (V)	I (A)	V/I
1	0.579	0.10	5.7900
2	0.978	0.17	5.7529
3	1.598	0.28	5.7071
4	1.976	0.34	5.8118
5	2.496	0.43	5.8047
6	2.953	0.51	5.7902
7	3.458	0.60	5.7633
8	4.068	0.71	5.7296
9	4.450	0.78	5.7051
10	4.917	0.86	5.7174
11	5.35	0.93	5.7527
12	5.75	1.01	5.6931
13	6.37	1.11	5.7387
14	6.60	1.15	5.7391

Fig. 8.2 Calculation of resistance for each pair of measured data

If you take a look to the third row of Table 8.2, you will understand that the value of resistor shows some changes with operating point. So, one question is that what is the best value to describe such a resistor. One way is to calculate the average of third row. According to Figs. 8.3 or 8.4 average value of third row is 5.7497.

Fig. 8.3 Calculation of average resistance (first method)

```
Command Window                                                                    ⊙
 >> V=[0.579,0.978,1.598,1.976,2.496,2.953,3.458,4.068,4.450,4.917,5.35,5.75,6.37,6.60];
 >> I=[0.10,0.17,0.28,0.34,0.43,0.51,0.60,0.71,0.78,0.86,0.93,1.01,1.11,1.15];
 >> V./I

 ans =

   Columns 1 through 8

     5.7900     5.7529     5.7071     5.8118     5.8047     5.7902     5.7633     5.7296

   Columns 9 through 14

     5.7051     5.7174     5.7527     5.6931     5.7387     5.7391

 >> sum(ans)/length(ans)

 ans =

     5.7497

fx >>
```

Fig. 8.4 Calculation of average resistance (second method)

Another and better way to solve this problem is to use curve fitting. We want to find value of G and b such that the function (curve) $i = Gv + b$ minimizes the following cost function:

$$\sum_{n=1}^{14}(i - i_n)^2 = \sum_{n=1}^{14}(Gv_n + b - i_n)^2 \tag{8.1}$$

You can define your curve as $i = Gv$ and minimizes the following cost function as well:

$$\sum_{n=1}^{14}(i - i_n)^2 = \sum_{n=1}^{14}(Gv_n - i_n)^2 \tag{8.2}$$

Let's continue with the first option, i.e. $i = Gv + b$. Enter the data to MATLAB and run the curve fitting toolbox with the aid of cftool command (Fig. 8.5). Curve fitting toolbox is shown in Fig. 8.6.

```
Command Window                                                                        ⊙
  >> V=[0.579,0.978,1.598,1.976,2.496,2.953,3.458,4.068,4.450,4.917,5.35,5.75,6.37,6.60];
  >> I=[0.10,0.17,0.28,0.34,0.43,0.51,0.60,0.71,0.78,0.86,0.93,1.01,1.11,1.15];
fx >> cftool|
  <                                                                                   >
```

Fig. 8.5   Running the toolbox with cftool command

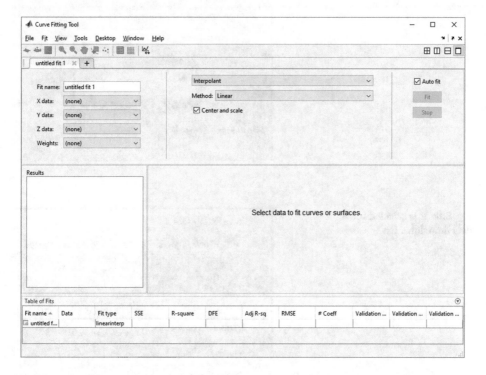

Fig. 8.6   Curve fitting toolbox

Select the V for X data (Figs. 8.7 and 8.8).

Fig. 8.7 X data drop down list

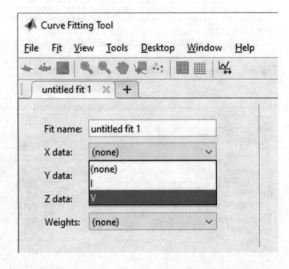

Fig. 8.8 V is selected for X data drop down list

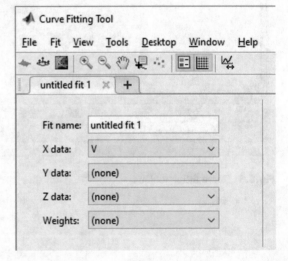

Select the I for Y data (Figs. 8.9 and 8.10).

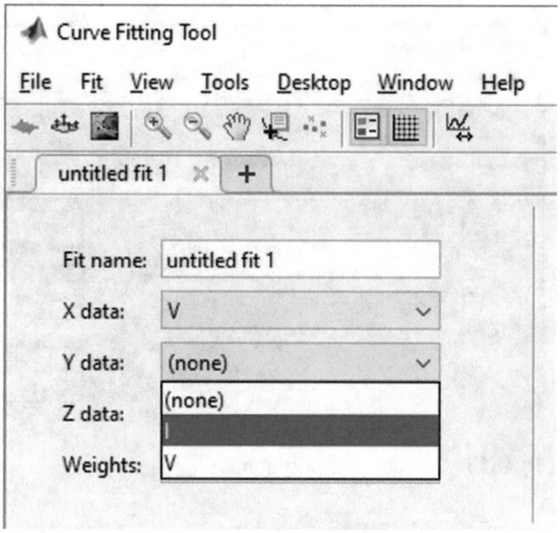

Fig. 8.9   Y data drop down list

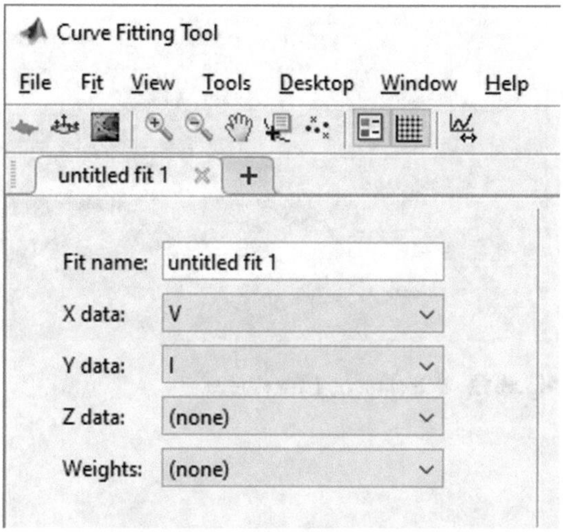

Fig. 8.10   I is selected for Y data drop down list

Select the polynomial (Fig. 8.11). Then select degree 1 (Fig. 8.12).

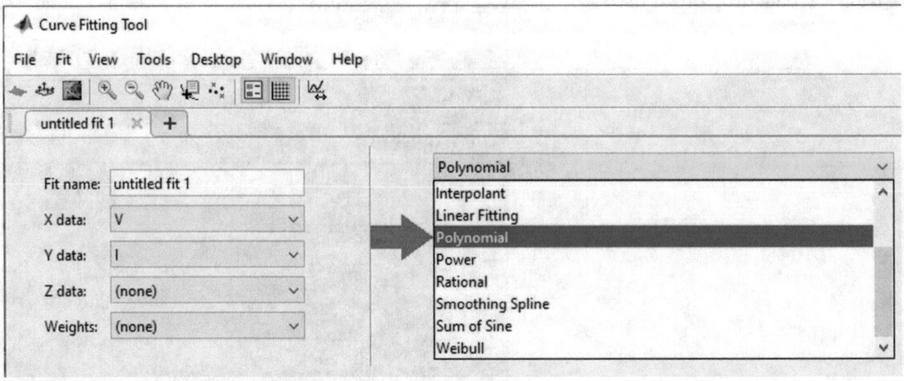

Fig. 8.11 Selection of polynomial

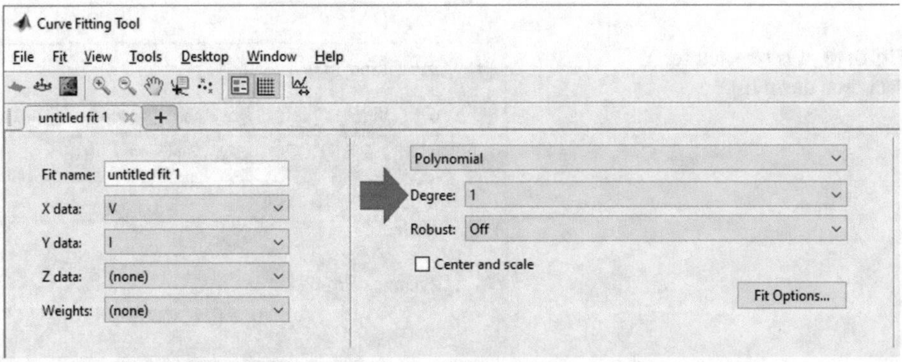

Fig. 8.12 Selection of 1 for degree

Ensure that auto fit is checked (Fig. 8.13). The toolbox finds the best line that pass through your data points and show it graphically. The toolbox also gives the equation of obtained curve to you as well.

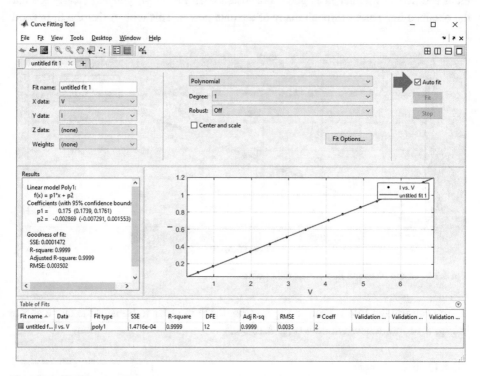

Fig. 8.13 Toolbox found the best line that pass through data points

8.3   Graphical Comparison of Estimation with Measured Data

The toolbox shows the graph of obtained function versus data points so you can compare
them and decide whether fit is good. For instance, in Fig. 8.14, you can clearly see that
the obtained line passes through many of the data points. This tells us that the fit is good.

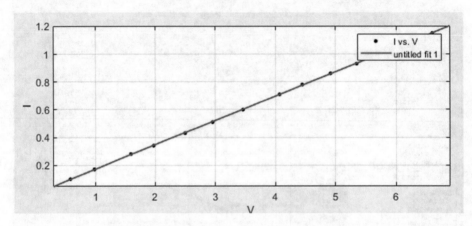

Fig. 8.14 Graph of found line versus data points

You can use the File > Print to Figure (Fig. 8.15) in order to transfer the curve shown
in Fig. 8.14 into a new window (Fig. 8.16).

Fig. 8.15 File > Print to
Figure

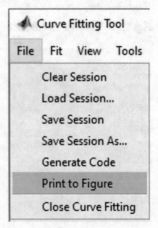

Fig. 8.16 Graph is exported
into the MATLAB environment

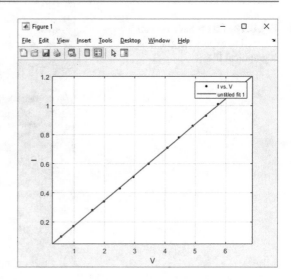

You can add a title to the graph with the aid of Insert > Title (Fig. 8.17). Insert > X
Label or Insert > Y Label can be used to add or edit the X axis or Y axis labels.

Fig. 8.17 Addition of title to
the graph

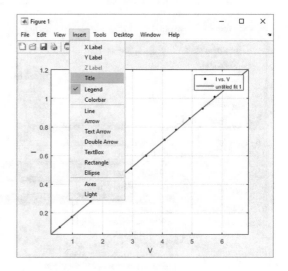

You can use the File > Save as in order to export the graph as a graphical file (Figs. 8.18
and 8.19).

Fig. 8.18 File > Save As can be used to export the graph as a graphical file

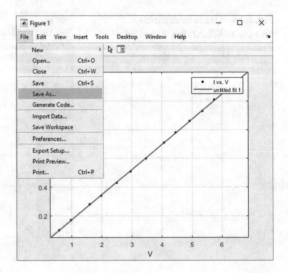

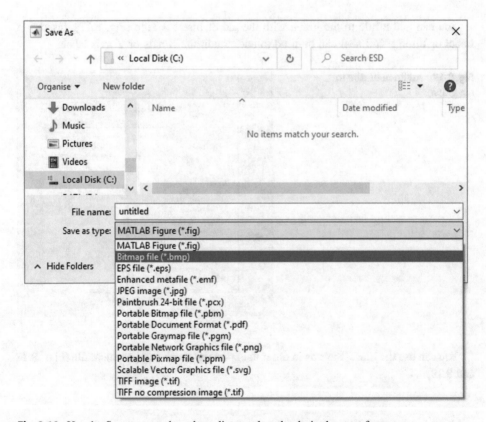

Fig. 8.19 Use the Save as type drop down list to select the desired output format

8.4 Quantitative Comparison of Estimation with Data

The toolbox shows the equation of the found curve in the bottom left corner of the screen (Fig. 8.20).

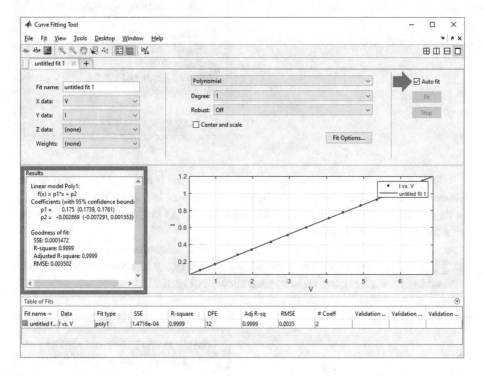

Fig. 8.20 Results section of the toolbox

Obtained results are shown in Fig. 8.21. The best line which pass through our data is $f(x) = 0.175x - 0.002869$. Note that R-square is quite close to 1. When R-square is close to one, you fit is good and reliable. When value of R-square become close to zero, then your fit is not good and reliable. Value of R-square is between zero and one.

Fig. 8.21 Equation of
obtained line is shown in the
results section

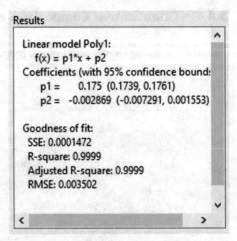

Results

Linear model Poly1:
    f(x) = p1*x + p2
Coefficients (with 95% confidence bound:
    p1 =   0.175 (0.1739, 0.1761)
    p2 =   -0.002869 (-0.007291, 0.001553)

Goodness of fit:
SSE: 0.0001472
R-square: 0.9999
Adjusted R-square: 0.9999
RMSE: 0.003502

In this example, Y axis is assigned to current and x axis is assigned to voltage, we can
deduce that the best line that pass through our data points is $i = 0.175v - 0.002869$. If
we neglect the -0.002869 term, then we reach to $i = 0.175v$. In other words, the best
value for resistance R1 is $\frac{1}{0.175} = 5.7143\,\Omega$.

According to Fig. 8.21, the SSE is 0.0001472. SSE stands for Sum of Square Errors,
i.e., $\sum_{n=1}^{14}(p_1.v_n + p_2 - i_n)^2$. The commands shown in Fig. 8.22 compute the SSE. Value
obtained in Fig. 8.22 is quite close to what shown in Fig. 8.21.

Fig. 8.22 Calculation of SSE

Command Window

```
>> Iest=0.175*V-0.002869;
>> sum((I-Iest)*(I-Iest)')

ans =

    1.4716e-04

fx >>
```

According to Fig. 8.21, the RMSE is 0.003502. RMSE stands for Root Mean Square Error, i.e., $\sqrt{\frac{1}{14}\sum_{n=1}^{14}(p_1.v_n + p_2 - i_n)^2}$. The commands shown in Fig. 8.23 compute the RMSE. Value obtained in Fig. 8.23 is quite close to what shown in Fig. 8.21.

```
Command Window                                                    ⊙
>> Iest=0.175*V-0.002869;
>> sqrt(sum((I-Iest)*(I-Iest)')/length(I))

ans =

     0.0032

f𝑥 >> |
```

Fig. 8.23 Calculation of RMSE

8.5   Example 2: Nonlinear Curve Fitting

Consider the simple circuit shown in Fig. 8.24. This voltmeter and ammeter measures the voltage and current of the diode. Measured voltages and currents are shown in Table 8.3.

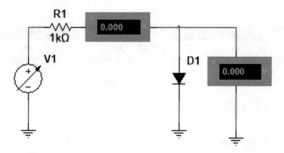

Fig. 8.24 Simple circuit to measure voltage and current of a diode

Table 8.3 I–V data for circuit shown in Fig. 8.24

No.	Diode voltage (V)	Diode current (mA)
1	0.50	0.50
2	0.53	0.97
3	0.55	1.44
4	0.57	1.93
5	0.58	2.42
6	0.60	3.40
7	0.61	4.38
8	0.62	5.37
9	0.63	6.36
10	0.64	7.34
11	0.68	17.00
12	0.70	24.00
13	0.72	34.00
14	0.73	39.00

We want find an equation with for this data. Let's assume that the relation between current and voltage is $i = a(e^{b.v} - 1)$. We want to find best value for a and b. Enter the data in Table 8.3 to MATLAB (Fig. 8.25).

```
Command Window                                                                    ⊙
 >> V=[0.5,0.53,0.55,0.57,0.58,0.6,0.61,0.62,0.63,0.64,0.68,0.70,0.72,0.73];
 >> I=1e-3*[0.50,0.97,1.44,1.93,2.42,3.40,4.38,5.37,6.36,7.34,17.00,24.00,34.00,39.00];
fx >> |
```

Fig. 8.25 Entering the data into MATLAB

Run the toolbox (Figs. 8.26 and 8.27).

```
Command Window                                                                    ⊙
 >> V=[0.5,0.53,0.55,0.57,0.58,0.6,0.61,0.62,0.63,0.64,0.68,0.70,0.72,0.73];
 >> I=1e-3*[0.50,0.97,1.44,1.93,2.42,3.40,4.38,5.37,6.36,7.34,17.00,24.00,34.00,39.00];
fx >> cftool|
```

Fig. 8.26 Running the toolbox with cftool command

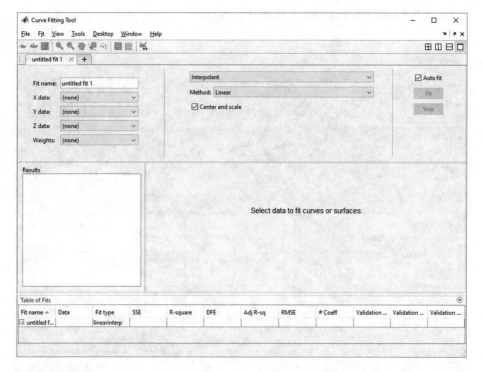

Fig. 8.27   Curve fitting toolbox is run

Select V for X Data and I for Y Data (Fig. 8.28).

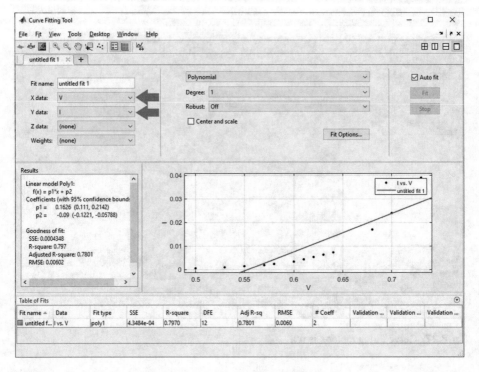

Fig. 8.28 X data and Y data boxes are filled with V and I, respectively

Select the custom equation (Fig. 8.29).

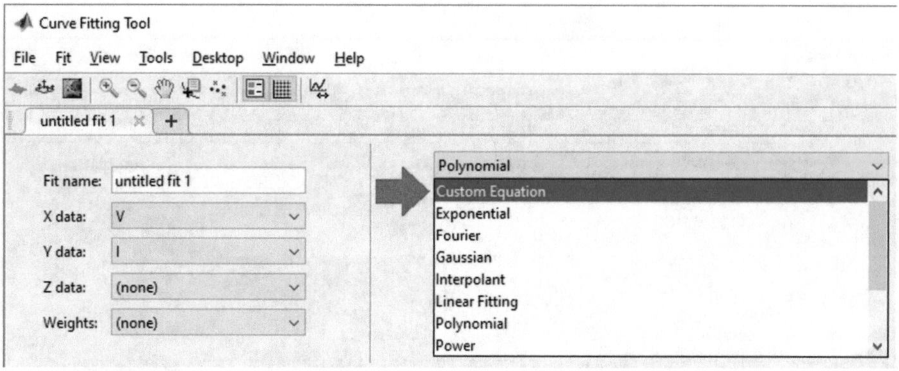

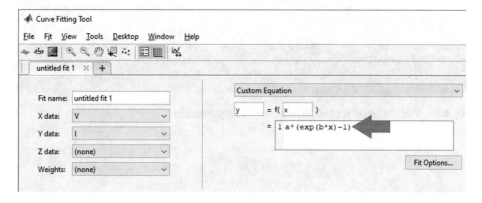

Fig. 8.29 Custom equation is selected

Enter the equation (Fig. 8.30).

Fig. 8.30 Desired equation is entered into the toolbox

Ensure that auto fit is checked (Fig. 8.31). Best values for a and b will be shown in the Results section in left bottom corner of the screen.

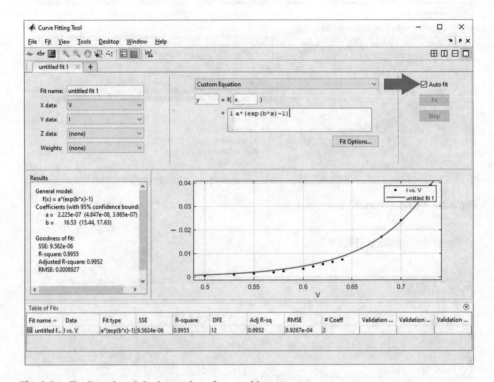

Fig. 8.31 Toolbox found the best values for a and b parameters

The obtained graph pass close to data points (Fig. 8.32). So we expect a good fit and we expect the R-square to be close to one.

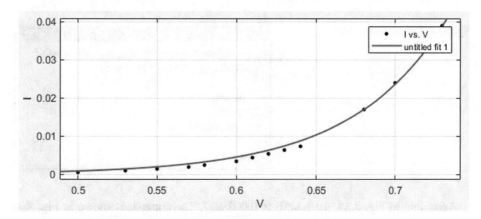

Fig. 8.32 Graph of found curve versus data points

According to the result section, best value for a is 2.225×10^{-7} and best value for b is 16.53 (Fig. 8.33). So, the best equation for our data is $I = 2.225 \times 10^{-7} \left(e^{16.53V} - 1 \right)$ where I and V are diode current and voltage, respectively. I is in Amps and V is in Volts.

Fig. 8.33 Best values for a and b parameters are shown in the results section

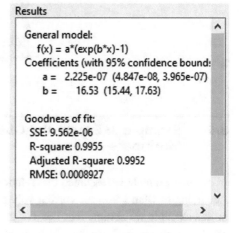

According to Fig. 8.33, the SSE is 9.562×10^{-6}. The commands shown in Fig. 8.34 compute the SSE, i.e., $\sum_{n=1}^{14}(2.225\times10^{-7}(e^{16.53v_n}-1)-i_n)^2$. Value obtained in Fig. 8.34 is quite close to what shown in Fig. 8.33.

Fig. 8.34 Calculation of SSE

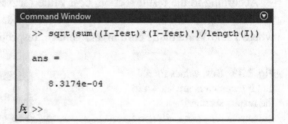

```
Command Window                                        ⊙

    >> Iest=2.225e-7*(exp(16.53*V)-1);
    >> sum((I-Iest)*(I-Iest)')


    ans =


       9.6851e-06

fx >>
```

According to Fig. 8.33, the RMSE is 0.0008927. The commands shown in Fig. 8.35 compute the RMSE, i.e., $\sqrt{\frac{1}{14}\sum_{n=1}^{14}(2.225\times10^{-7}(e^{16.53v_n}-1)-i_n)^2}$. Value obtained in Fig. 8.35 is quite close to what shown in Fig. 8.33.

Fig. 8.35 Calculation of RMSE

```
Command Window                                        ⊙

    >> sqrt(sum((I-Iest)*(I-Iest)')/length(I))

    ans =


       8.3174e-04

fx >>
```

8.6 Example 3: Export the Obtained Equation to the MATLAB Workspace

In our first example we assumed that relation between current and voltage is $i = Gv + b$. According to Ohm's law, the current voltage relationship for a conductor is $i = Gv$. So, in this example we assume that $i = Gv$ and we want to find the best value for G. Data points are shown in Table 8.1. Settings of curve fitting toolbox is shown in Fig. 8.36. Results section is shown in Fig. 8.37 as well. So, the best equation for this data set is $i = 0.1744 \times v$. So, best value for the resistor is $\frac{1}{0.1744} = 5.74\,\Omega$.

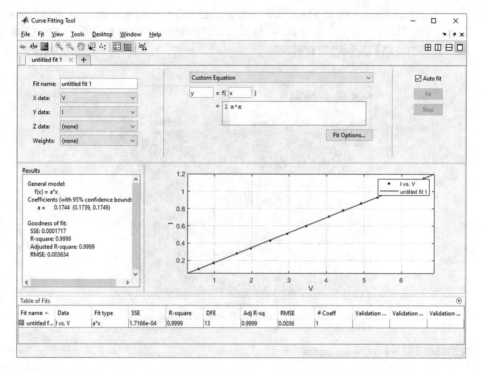

Fig. 8.36 Obtaining the best line with equation of $i = Gv$ for data shown in Table 8.1

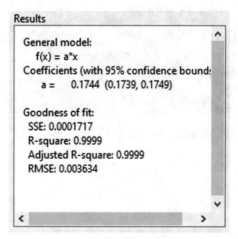

Fig. 8.37 Best value for parameter G is found

Let's export the model to MATLAB environment. Click the Fit > Save to Workspace (Fig. 8.38). This opens the Save Fit to MATLAB Workspace window (Fig. 8.39). Click the OK button in Fig. 8.39 to proceed.

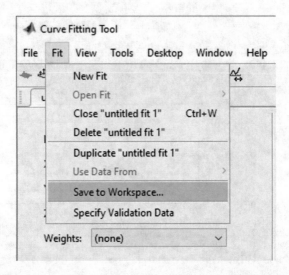

Fig. 8.38 File > Save to workspace

Fig. 8.39 Save fit to MATLAB workspace window

Now minimize the toolbox and return to MATLAB environment. As shown in Fig. 8.40, new variables are added to the Workspace.

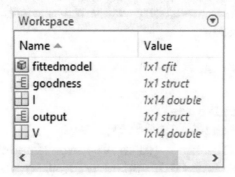

Fig. 8.40 New variables are imported into the MATLAB workspace

Observe the content of the fittedmodel variable (Fig. 8.41).

```
Command Window
>> fittedmodel

fittedmodel =

      General model:
      fittedmodel(x) = a*x
      Coefficients (with 95% confidence bounds):
         a =       0.1744   (0.1739, 0.1749)
fx >>
```

Fig. 8.41 Contents of fittedmodel variable

You can use the dot operator in order to obtain the coefficients (Fig. 8.42).

```
Command Window                                              ⏷

   >> fittedmodel

fittedmodel =

     General model:
     fittedmodel(x) = a*x
     Coefficients (with 95% confidence bounds):
       a =        0.1744  (0.1739, 0.1749)
   >> Resistance= 1/ fittedmodel.a

Resistance =

     5.7339

fx >> |
```

Fig. 8.42 Dot operator permits you to obtain access to the calculated model parameters

Exercises

1. Find the best second order polynomial which fits to the data given in Table 8.4.
2. Write suitable MATLAB code to compute the SSE and RMSE of model obtained in exercise 1. Compare output of your code with values given by toolbox.

Table 8.4 Data for exercise 1

x	y
1	1
2	5.3
3	10.5
4	16.8
5	25
6	36.9
7	50.8

References for Further Study

1. Numerical Analysis, Burden, RL., Fairs D., Burden AM., Cengage Learning, 2015.
2. Numerical Methods: Using MATLAB, Lindfield G., Penny J., Academic Press, 2018.
3. Applied Numerical Methods with MATLAB for Engineers and Scientists, Chapra S., McGraw Hill, 2005.

Drawing Graphs with MATLAB® 9

9.1 Introduction

In this chapter you will learn how to draw different types of graphs with MATLAB. You will learn how to draw the graph of functions of one variables, graph of functions of two variables, graph of functions of three variables and different types of statistical graphs with MATLAB.

9.2 fplot Command

Assume that you want to draw the graph of $\sin(x)$ for $[0, 2\pi]$ interval. The commands shown in Fig. 9.1 do this job for you. Output of this code is shown in Fig. 9.2.

Fig. 9.1 fplot command can be used to draw symbolic expressions

```
Command Window
>> syms x
>> fplot(sin(x),[0,2*pi])
>> grid on
fx >>
```

© The Author(s), under exclusive license to Springer Nature Switzerland AG 2023
F. Asadi, *Applied Numerical Analysis with MATLAB®/Simulink®*,
Synthesis Lectures on Engineering, Science, and Technology,
https://doi.org/10.1007/978-3-031-19366-8_9

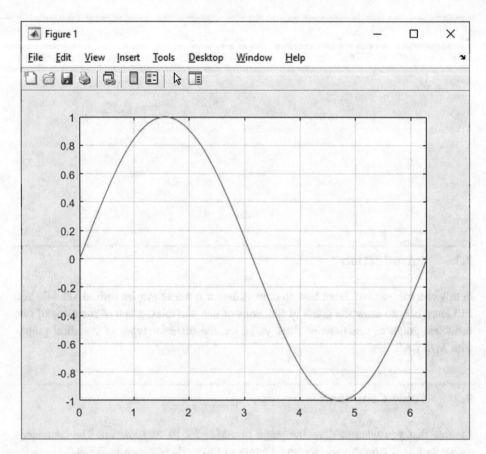

Fig. 9.2 Output of code in Fig. 9.1

You click on any point in order to read its coordinate (Fig. 9.3).

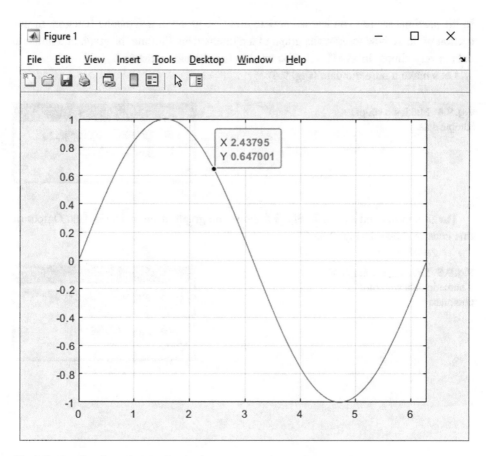

Fig. 9.3 Reading the points on the graph

9.3   Plotting the Graph of a Numeric Data

In the previous section you learned how to draw the graph of a symbolic function. In this section we learn how to draw the graph of a numeric data. Plotting the graph of a numeric data is very simple in MATLAB. You need to use the plot command.

Let's make a numeric data (Fig. 9.4).

Fig. 9.4 Making a simple
sample data

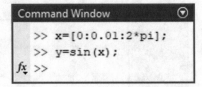

The plot command shown in Fig. 9.5 draws the graph of the numeric data. Output of this code is shown in Fig. 9.6.

Fig. 9.5 Drawing the graph of
a numeric data with plot
command

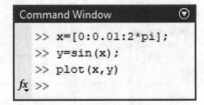

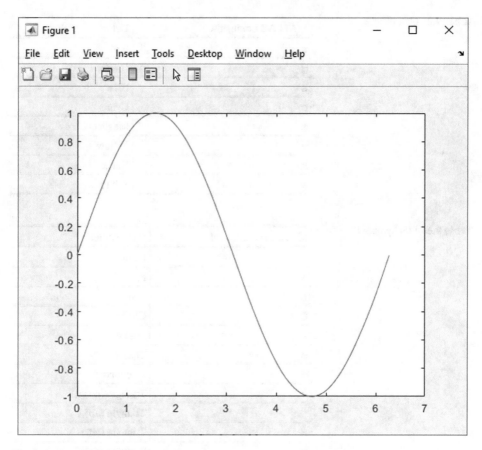

Fig. 9.6 Output of code in Fig. 9.5

You can add the operators shown in Tables 9.1, 9.2 and 9.3 to produce more custom plots.

Table 9.1 Types of lines

MATLAB command	Type of line
-	Solid
:	Dotted
-.	Dashdot
--	Dashed

Table 9.2 Colors

MATLAB command	Color
r	Red
g	Green
b	Blue
c	Cyan
m	Magenta
y	Yellow
k	Black
w	White

Table 9.3 Plot symbols

MATLAB command	Plot symbol
.	Point
+	Plus
*	Star
o	Circle
x	x-mark
s	Square
d	Diamond
v	Triangle (down)
^	Triangle (up)
<	Triangle (left)
>	Triangle (right)

Let's study an example. Values of voltage and current for a resistor is shown in Table 9.4. We want to plot the graph of this data. We want to show the data points with circles and connect them together using dashed line with black color. The vertical axis and horizontal axis must have the labels "Current (A)" and "Voltage (V)", respectively. The title of the graph must be "I–V for a resistor". The commands shown in Fig. 9.7 do what we need. The result is shown in Fig. 9.8.

Table 9.4 V–I values for resistor R1

V (V)	I (A)
0.499	0.10
0.985	0.20
1.508	0.31
1.969	0.41
2.528	0.53
2.935	0.61
3.481	0.73
3.971	0.83
4.486	0.94
4.960	1.04
5.502	1.15
6.007	1.26
6.60	1.38

```
Command Window
>> V=[0.499 0.985 1.508 1.969 2.528 2.935 3.481 3.971 4.486 4.960 5.502 6.007 6.60];
>> I=[0.1 0.2 0.31 0.41 0.53 0.61 0.73 0.83 0.94 1.04 1.15 1.26 1.38];
>> plot(V,I,'--ko')
>> title('I-V for a resistor')
>> xlabel('Voltage(V)')
>> ylabel('Current(A)')
>> grid on
fx >>
```

Fig. 9.7 Drawing the graph of data in Table 9.4

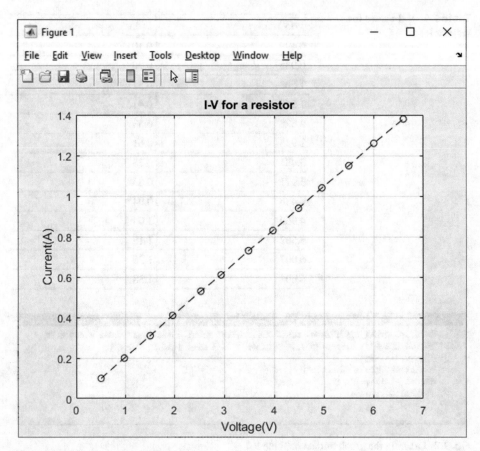

Fig. 9.8 Output of code in Fig. 9.7

The commands shown in Fig. 9.9 draw the I–V graph of Table 9.4. However, it uses red star for data points and solid black color for connecting the data points together (Fig. 9.10).

```
Command Window
  >> V=[0.499 0.985 1.508 1.969 2.528 2.935 3.481 3.971 4.486 4.960 5.502 6.007 6.60];
  >> I=[0.1 0.2 0.31 0.41 0.53 0.61 0.73 0.83 0.94 1.04 1.15 1.26 1.38];
  >> plot(V,I,'k',V,I,'r*'),grid minor
fx >>
```

Fig. 9.9 Drawing the graph of data in Table 9.4

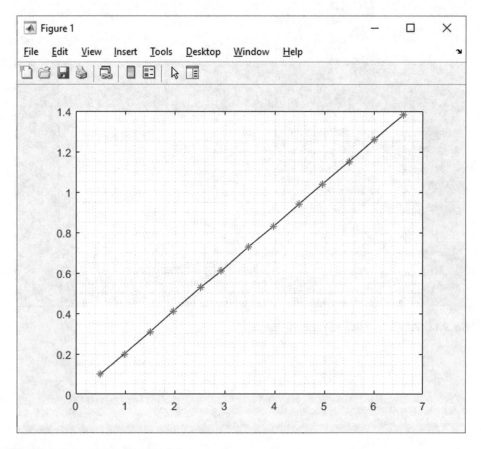

Fig. 9.10 Output of code in Fig. 9.9

9.4   Addition of Labels and Title to the Drawn Graph

Figure 9.10 doesn't have any labels and title. You can add the desired labels and titles to it with the aid of insert menu (Fig. 9.11).

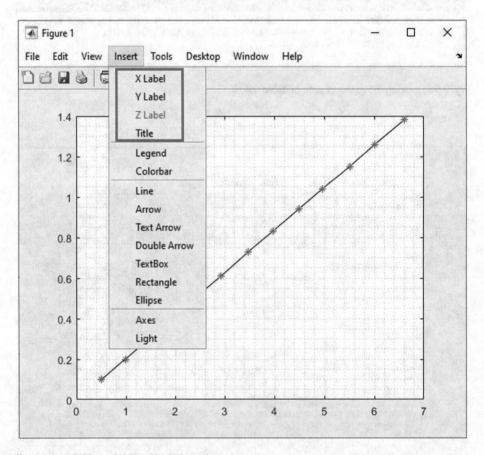

Fig. 9.11 Addition of title and labels to the axis

9.5   Exporting the Drawn Graph as a Graphical File

You can copy the drawn graph to the clipboard easily with the aid of Edit > Copy Figure (Fig. 9.12). After copying the graph to the clipboard you can easily paste it in programs like MS Word® by pressing Ctrl + V.

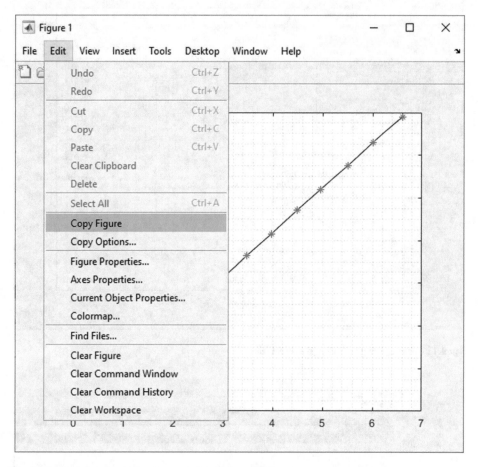

Fig. 9.12 Copying the drawn figure to the clipboard memory

You can save the drawn graph as a graphical file as well. To do this, click use the File > Save As (Fig. 9.13). After clicking, the save as window appears. Select the desired output format from the Save as type drop down list (Fig. 9.14).

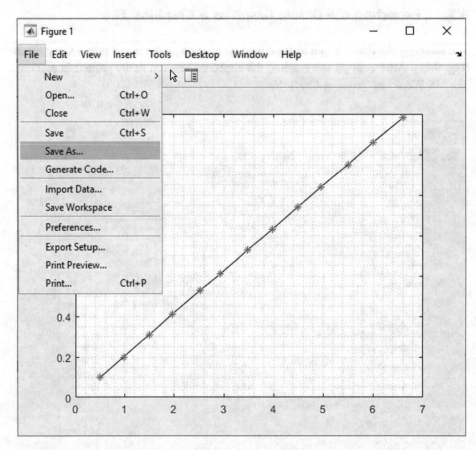

Fig. 9.13 Saving the graph as a graphical file

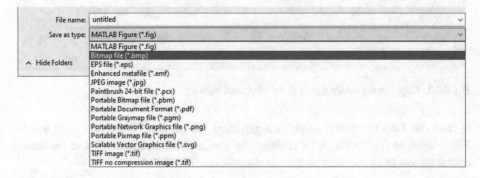

Fig. 9.14 Selection of desired type of file

9.6   Drawing Two or More Graphs on the Same Axis

Sometimes you need to show two or more datasets on the same graph. You need to use the hold on command to show two or more graphs simultaneously. Assume that we have another dataset (Table 9.5) and we want to show both datasets (Tables 9.4 and 9.5) on the same graph.

Table 9.5 V–I values for resistor R2

V (V)	I (A)
0.579	0.10
0.978	0.17
1.598	0.28
1.976	0.34
2.496	0.43
2.953	0.51
3.458	0.60
4.068	0.71
4.450	0.78
4.917	0.86
5.35	0.93
5.75	1.01
6.37	1.11
6.60	1.15

The commands shown in Fig. 9.15 draws the graph of both datasets on the same graph. Output of this code is shown in Fig. 9.16.

```
Command Window
>> V1=[0.499 0.985 1.508 1.969 2.528 2.935 3.481 3.971 4.486 4.960 5.502 6.007 6.60];
>> I1=[0.1 0.2 0.31 0.41 0.53 0.61 0.73 0.83 0.94 1.04 1.15 1.26 1.38];
>> V2=[0.579 0.978 1.598 1.976 2.496 2.953 3.458 4.068 4.450 4.917 5.35 5.75 6.37 6.60];
>> I2=[0.1 0.17 0.28 0.34 0.43 0.51 0.60 0.71 0.78 0.86 0.93 1.01 1.11 1.15];
>> plot(V1,I1,'b',V1,I1,'r*')
>> hold on
>> plot(V2,I2,'k',V2,I2,'r+')
>> grid minor
>> xlabel('Voltage (V)')
>> ylabel('Current (A)')
>> title('Comparison of I-V plot of R1 and R2')
fx >>
```

Fig. 9.15 Drawing the graph of Tables 9.4 and 9.5 on the same graph

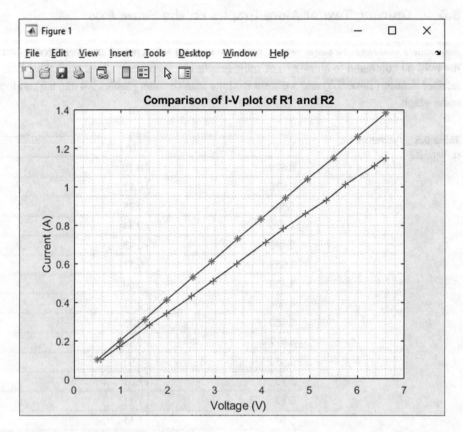

Fig. 9.16 Output of code in Fig. 9.15

You can use the Insert > Legend to show which graph belongs to which resistor (Fig. 9.17).

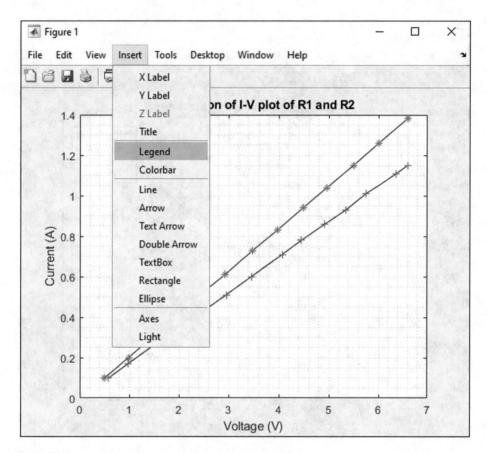

Fig. 9.17   Insert > Legend can be used to add a legend to the graph

After clicking the Insert > Legend, the legend shown in Fig. 9.18 will be added to the graph.

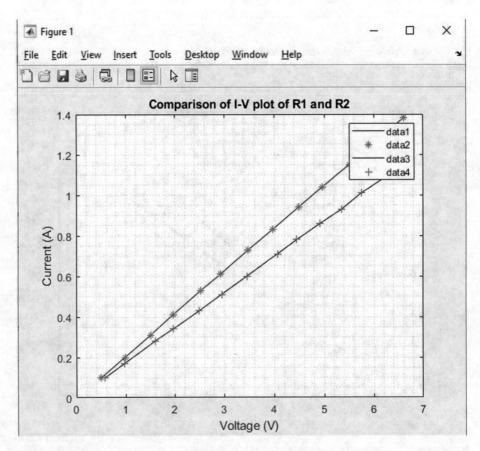

Fig. 9.18 Legend is added to the graph

Double click the data 1 in the legend box (Fig. 9.18) and enter the desired text. Repeat this for data 2, data 3 and data 4 in Fig. 9.18. You can move the legend box by clicking on it, holding down the mouse button and dragging it to the desired location (Fig. 9.19). You can even right click on the legend box and use the predefined locations (Fig. 9.20).

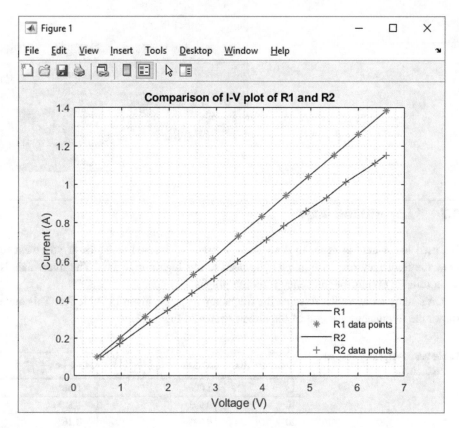

Fig. 9.19 Customized legend

Fig. 9.20 Default locations
for legend

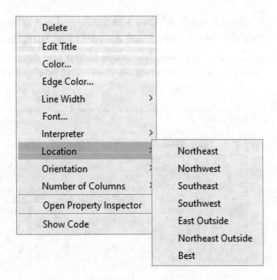

9.7 Logarithmic Axis

We used linear axis in order to draw the I-V graph of studied resistors. If you want to draw the frequency response graphs you need to use logarithmic axis. The linear axis is not a suitable option for frequency response graphs.

Let's study an example. Assume the frequency response given in Table 9.6. This table shows the frequency response of the circuit shown in Fig. 9.21.

Table 9.6 Frequency response of a RC circuit

Frequency (Hz)	Magnitude $\left(\left|\frac{V_o(j\omega)}{V_{in}(j\omega)}\right|\right)$	Phase $\left(\sphericalangle \frac{V_o(j\omega)}{V_{in}(j\omega)}\right)$
1	1.000	$-0.36°$
10	0.998	$-3.60°$
20	0.992	$-7.16°$
50	0.954	$-17.44°$
100	0.847	$-32.13°$
150	0.728	$-43.30°$
200	0.623	$-51.48°$
250	0.537	$-57.51°$
300	0.469	$-62.05°$
350	0.414	$-65.54°$
400	0.370	$-68.30°$
450	0.333	$-70.51°$
500	0.303	$-72.34°$
550	0.278	$-73.85°$
600	0.256	$-75.14°$

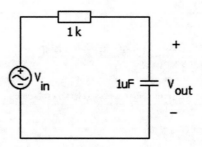

Fig. 9.21 Simple RC circuit

The commands shown in Fig. 9.22, draws the frequency response of the data in Table 9.6. The command semilogx is used to draw the frequency response graph. Output of this code is shown in Fig. 9.23.

```
Command Window
>> f=[1 10 20 50 100 150 200 250 300 350 400 450 500 550 600];
>> Amp=[1 0.998 0.992 0.954 0.847 0.728 0.623 0.537 0.469 0.414 0.370 0.333 0.303 0.278 0.256];
>> Phase=-[0.36 3.6 7.16 17.44 32.13 43.30 51.48 57.51 62.05 65.54 68.30 70.51 72.34 73.85 75.14];
>> subplot(211),semilogx(f,20*log10(Amp)),grid minor
>> title('Frequency responce of the RC circuit')
>> xlabel('Freq(Hz.)')
>> ylabel('Amplitude (dB)')
>> subplot(212),semilogx(f,Phase),grid minor
>> xlabel('Freq(Hz.)')
>> ylabel('Phase(Degrees)')
fx >> |
```

Fig. 9.22 Drawing the graph of data in Table 9.6

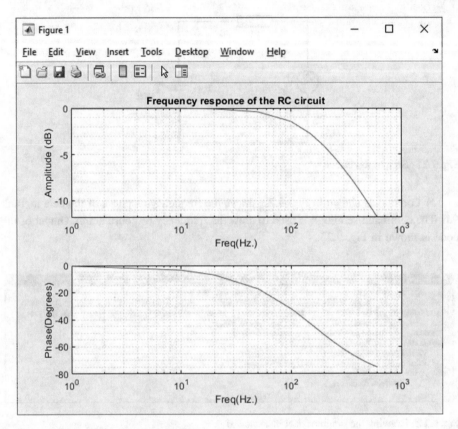

Fig. 9.23 Output of code in Fig. 9.22

9.8   Drawing 2D and 3D Parametric Graphs

You can use the plot3 command in order to draw 3D graphs. For instance, the code shown in Fig. 9.24 draws the plot of $\begin{cases} x = \cos(t) \\ y = \sin(t) \\ z = t \end{cases}$ for $0 \leq t \leq 2\pi$. Output of this code is shown in Fig. 9.25.

Fig. 9.24 plot3 is used to draw a parametric graph

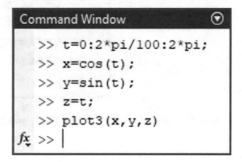

Fig. 9.25 Output of the code in Fig. 9.24

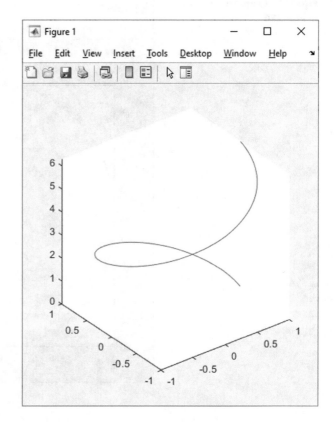

You can use the plot command in order to draw the 2D parametric graphs. For instance, the code in Fig. 9.26 draws the graph of $\begin{cases} x(t) = \cos(t) \\ y(t) = \sin(t) \end{cases}$ for $0 \le t \le 2\pi$. Output of this code is shown in Fig. 9.27.

```
Command Window                        ⊙
  >> t=0:2*pi/100:2*pi;
  >> x=cos(t);
  >> y=sin(t);
  >> plot(x,y)
fx >> |
```

Fig. 9.26 Drawing a 2D parametric graph with plot command

Fig. 9.27 Output of the code in Fig. 9.26

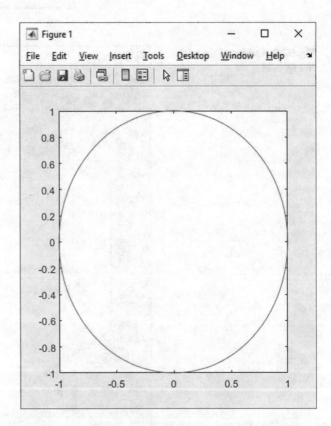

9.9 Polar Plot

Polar plots can be drawn in MATLAB easily with the aid of polarplot command. For instance, the commands shown in Fig. 9.28 draws the graph of $\rho = \sin(2\theta).\cos(2\theta)$ for $0 \leq \theta \leq 2\pi$. Output of the code is shown in Fig. 9.29. You can click on any point to read its's coordinate (Fig. 9.30).

```
Command Window                               ⊙

>> theta = 0:0.01:2*pi;
>> rho = sin(2*theta).*cos(2*theta);
>> polarplot(theta,rho)
fx >> |
```

Fig. 9.28 Drawing a polar plot with polarplot command

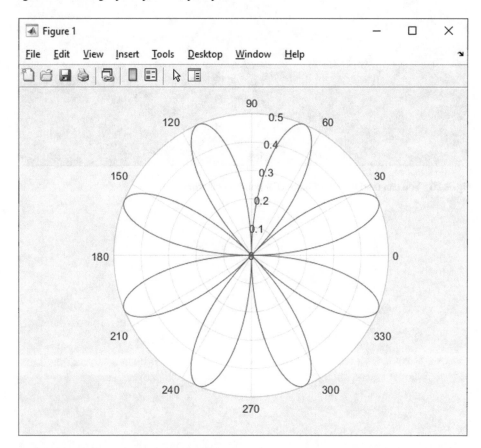

Fig. 9.29 Output of the code in Fig. 9.28

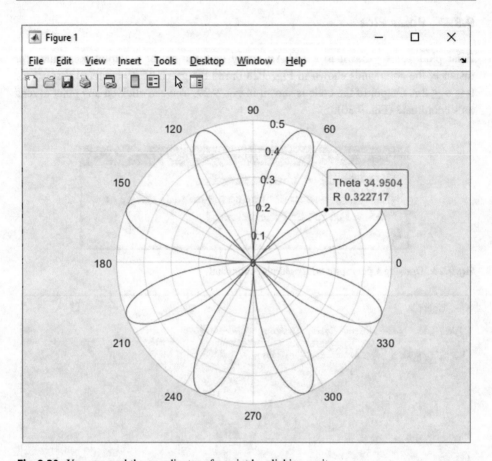

Fig. 9.30 You can read the coordinates of a point by clicking on it

9.10   3D Surfaces

You can draw 3D surfaces with meshgrid and surf commands easily. For instance, the commands shown in Fig. 9.31 draws the surface $z = \sin(x) + \cos(y)$ for $1 \leq x \leq 10$ and $1 \leq y \leq 20$. Output of the code is shown in Fig. 9.32.

```
Command Window                                          ⊙
   >> [X,Y] = meshgrid(1:0.1:10,1:0.1:20);
   >> Z = sin(X) + cos(Y);
   >> surf(X,Y,Z)
fx >>
```

Fig. 9.31   Drawing the surface $z = \sin(x) + \cos(y)$ for $1 \leq x \leq 10$ and $1 \leq y \leq 20$

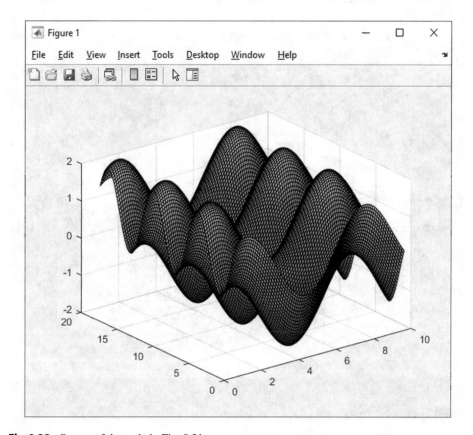

Fig. 9.32   Output of the code in Fig. 9.31

Let's study another example. For instance, let's draw the $z = 100(x_2 - x_1^2)^2 + (1 - x_1)^2$ for $-2 \leq x \leq 2$ and $-1 \leq y \leq 2$. The code shown in Fig. 9.33 do this job for us. Output of this code is shown in Fig. 9.34.

```
Command Window                                              ⦿
>> [X,Y]=meshgrid(-2:0.1:2,-1:0.1:2);
>> Z=100*(Y-X.^2).^2+(1-X).^2;
>> surf(X,Y,Z)
fx >> |
```

Fig. 9.33 Drawing the surface $z = 100(x_2 - x_1^2)^2 + (1 - x_1)^2$ for $-2 \leq x \leq 2$ and $-1 \leq y \leq 2$

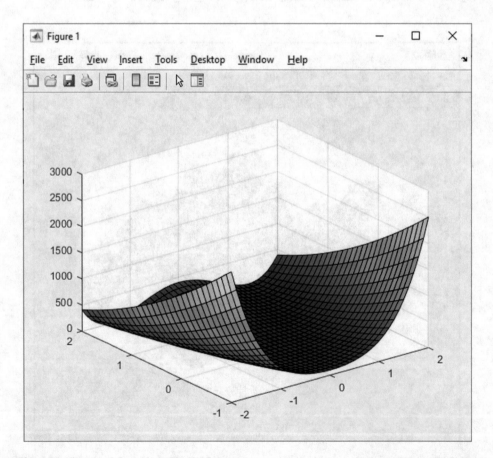

Fig. 9.34 Output of the code in Fig. 9.33

9.11   Pie Chart

Pie charts can be drawn with the aid of pie commands. For instance, Table 9.7 shows number of products sold by different sales personnel during one month.

Table 9.7 Number of sold products

Name	Number of sold products
John	100
Mary	180
Ted	52
Smith	130
Sarah	48

The code shown in Fig. 9.35 draws the pie chart of Table 9.7. Output of this code is shown in Fig. 9.36. In Fig. 9.37 legend is placed in the right top corner of the window.

```
Command Window
>> sales=[100 180 52 130 48];
>> labels={'john','mary','ted','smith','sarah'};
>> pie(sales)
>> lgd = legend(labels);
fx >> |
```

Fig. 9.35 Code for drawing the pie chart of Table 9.7

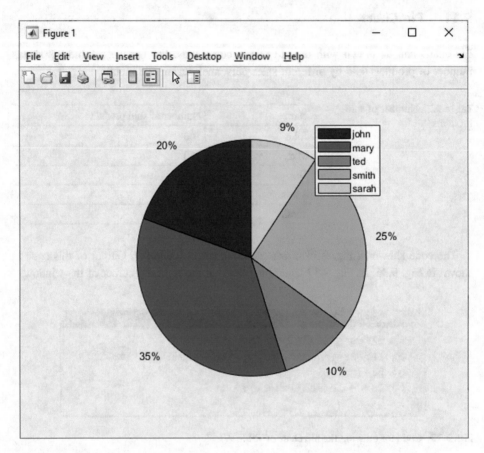

Fig. 9.36 Output of the code in Fig. 9.35

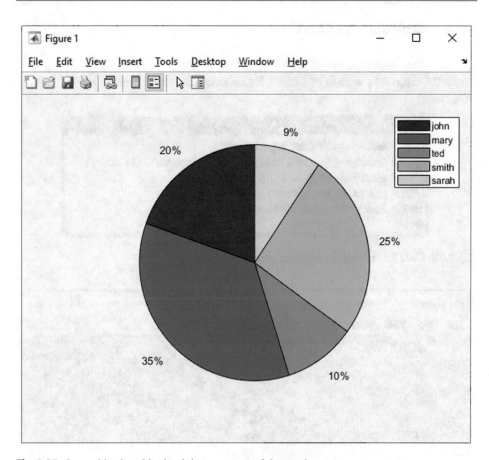

Fig. 9.37 Legend is placed in the right top corner of the graph

9.12   Exploded Pie Chart

You can draw exploded pie charts as well. For instance, the code shown in Fig. 9.38 draws the exploded pie chart of Table 9.7. Output of this code is shown in Fig. 9.39.

```
Command Window
>> sales=[100 180 52 130 48];
>> labels={'john','mary','ted','smith','sarah'};
>> explode=[1 1 1 1 1];
>> pie(sales,explode)
>> lgd = legend(labels);
fx >>
```

Fig. 9.38   Code for drawing the exploded pie chart of Table 9.7

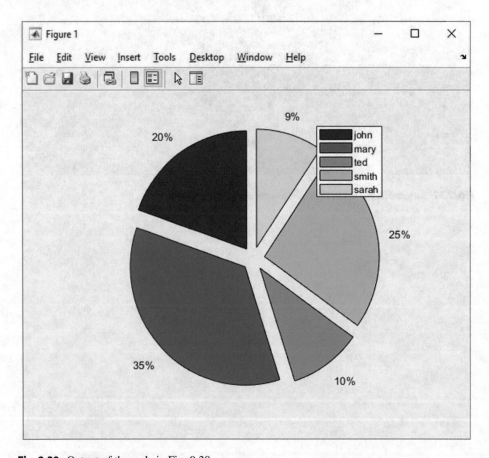

Fig. 9.39   Output of the code in Fig. 9.38

Now you can click on the legend and drag the legend in Fig. 9.39 to the location you want (Fig. 9.40).

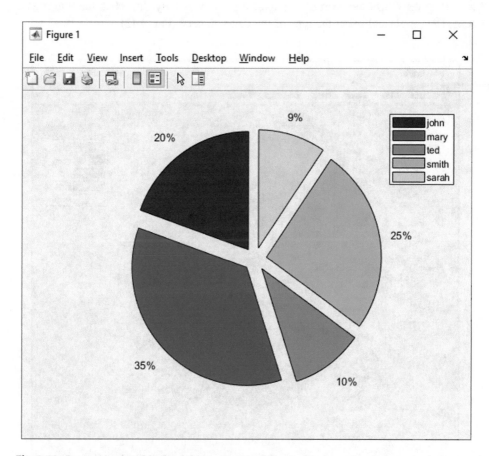

Fig. 9.40 Legend is placed in the right top corner of the graph

9.13   Export the Drawn Pie Chart as a Graphical File

Exporting the drawn pie chart as a graphical file is very easy. Just click the File > Save As… (Fig. 9.41) and select the type of file that you need (Fig. 9.42).

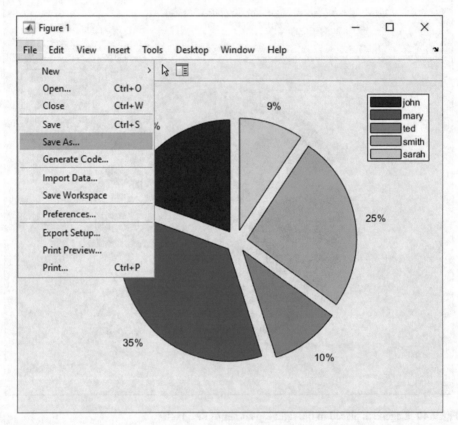

Fig. 9.41   File > Save As can be used to export the drawn pie chart as a graphical file

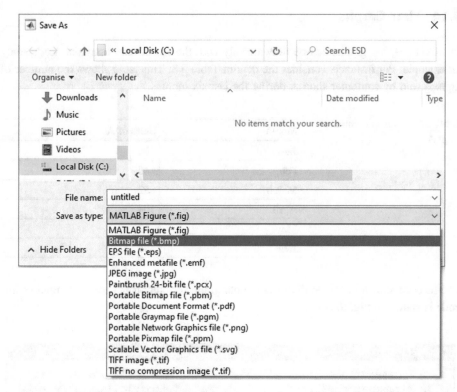

Fig. 9.42 Selecting the desired graphical format

9.14   Bar Graphs

In MATLAB, bar graphs can be drawn easily with the aid of bar command. Let's study an example. For instance, consider the data in Table 9.8. This table shows the number of laptops sold by computer shop A during the last six months.

Table 9.8 Sales of computer shop A	Month	Sales of A
	Jan	70
	Feb	40
	March	52
	April	104
	May	42
	June	56

The code shown in Fig. 9.43 draws the bar graph for computer shop A. Output of this code is shown in Fig. 9.44.

```
Command Window
>> X=categorical({'Jan','Feb','March','April','May','June'});
>> X=reordercats(X,{'Jan','Feb','March','April','May','June'});
>> SalesA=[70 40 52 104 42 56];
>> bar(X,SalesA)
>> title('Number of sold laptops')
fx >>
```

Fig. 9.43 Drawing the bar graph of Table 9.8

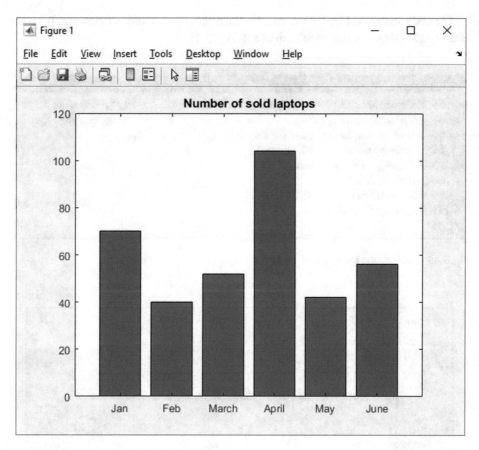

Fig. 9.44   Output of the code in Fig. 9.43

Let's study another example. In this example we want to compare sales of computer shop A with B. In other terms, sales of A and B must be drawn on the same graph. Table 9.9 shows the sales of these companies.

Table 9.9   Sales of computer shop A versus B

Month	Sales of A	Sales of B
Jan	70	60
Feb	40	50
March	52	53
April	104	80
May	42	60
June	56	70

The code shown in Fig. 9.45 draws the bar graph of computer shop A and B on the same graph. Output of this code is shown in Fig. 9.46.

```
Command Window                                                          ⊙
  >> X=categorical({'Jan','Feb','March','April','May','June'});
  >> X=reordercats(X,{'Jan','Feb','March','April','May','June'});
  >> SalesA=[70 40 52 104 42 56];
  >> SalesB=[60 50 53 80 60 70];
  >> Sales=[SalesA;SalesB];
  >> bar(X,Sales)
  >> title('Number of sold laptops')
  >> legend('A','B')
fx >>
```

Fig. 9.45 Drawing the bar graph of Table 9.9

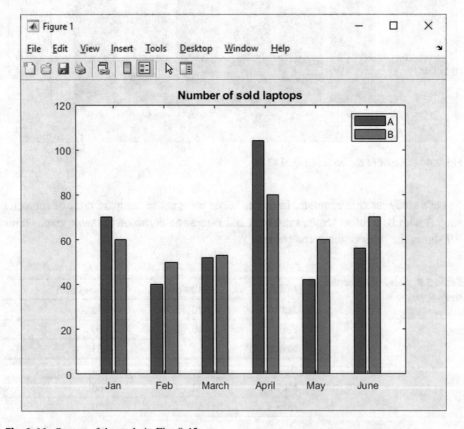

Fig. 9.46 Output of the code in Fig. 9.45

Exercises

1. Use fplot command to draw the graph of $e^{-x}\sin(3x)$. Use the graph to find its maximum value.

2. Use MATLAB to draw the graph of $\begin{cases} x = \cos(2t) \\ y = \sin(2t) \quad \text{for } 0 \le t \le 2\pi \\ z = 6t \end{cases}$

3. Use plot command to draw the graph of data shown in Table 9.10.

4. Use MATLAB to draw the surface of $z = (x_2^4 - x_1^2)^2 + (1 - x_1)^4$ for $-4 \le x \le 4$ and $-2 \le y \le 2$.

5. Sales of a company is shown in Table 9.11. Use MATLAB to draw the pie chart and bar graph of this table.

Table 9.10 Data for exercise 3

x	y
1	1
2	5.3
3	10.5
4	16.8
5	25
6	36.9
7	50.8

Table 9.11 Data for exercise 5

Month	Sales of A
Jan	300
Feb	450
March	620
April	600
May	450
June	800
July	960
Aug	600
Sep	300
Oct	890
Nov	450
Dec	280

References for Further Study

1. Chapman, S.: MATLAB Programming for Engineers, 6th edition, Cengage, 2019.
2. Hahn, B., Valentine, D.: Essential MATLAB for Engineers and Scientists, 7th edition, Academic Press, 2019.
3. Moore, H.: MATLAB for Engineers, 5th edition, Pearson, 2017.

MATLAB® Programming 10

10.1 Introduction

MATLAB permits you to write your own code beside the ready to use tools that it provides for you. In this chapter you will learn how to code some of the well-known numerical analysis algorithms in MATLAB environment. Being familiar with C programming is an advantage for this chapter.

10.2 MATLAB Editor

Command widow is not a suitable environment for writing long codes. It is better to use MATLAB editor to write long codes. You can activate the editor by typing the edit and press the enter key of the keyboard (Fig. 10.1). The MATLAB editor is shown in Fig. 10.2.

Fig. 10.1 Running the MATLAB editor

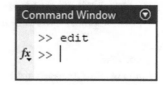

© The Author(s), under exclusive license to Springer Nature Switzerland AG 2023   257
F. Asadi, *Applied Numerical Analysis with MATLAB®/Simulink®*,
Synthesis Lectures on Engineering, Science, and Technology,
https://doi.org/10.1007/978-3-031-19366-8_10

Fig. 10.2 MATLAB editor

10.3   Simple Game

Let's study a simple program. In the following code, the user tries to guess a random number that is selected from 1 to 100. This code shows how to get input from the user, display suitable messages and how to use the if command.

```
clc
clear all

maximumTry=5;
maximumNumber=100;

target=floor(maximumNumber*rand()); %the random target number
guess=input('Please enter your guess:');

noOfTrial=0;
while (guess~=target)
noOfTrial=noOfTrial+1;

if (guess>target)
   if (guess<0 ||noOfTrial>maximumTry)
     break
   end
       guess=input('Your   guess   is   bigger   than   the   target   num-
ber. Enter new guess:');
```

```
else
    if (guess<0||noOfTrial>maximumTry)
        break
    end
        guess=input('Your  guess  is  smaller  than  the  target  num-
ber. Enter new guess:');
end

end

if (guess==target)
    disp('Congratulation you win!')
end
disp ('Game finished.')
```

Enter the code to the MATLAB editor (Fig. 10.3) and save it by pressing the Ctrl+S (Fig. 10.4). The MATLAB codes are saved with the.m extension. Note that MATLAB uses the blue and green color to show MATLAB language keywords and comments, respectively.

```
Editor - Untitled*                                                              ⊙ ×
  Untitled*  ×  +
 1      clc
 2      clear all
 3
 4      maximumTry=5;
 5      maximumNumber=100;
 6
 7      target=floor(maximumNumber*rand()); %the random target number
 8      guess=input('Please enter your guess:');
 9
10      noOfTrial=0;
11   ⊟ while (guess~=target)
12      noOfTrial=noOfTrial+1;
13
14      if (guess>target)
15          if (guess<0 ||noOfTrial>maximumTry)
16              break
17          end
18          guess=input('Your guess is bigger than the target number. Enter new guess:');
19      else
20          if (guess<0||noOfTrial>maximumTry)
21              break
22          end
```

Fig. 10.3 Code is entered to MATLAB editor

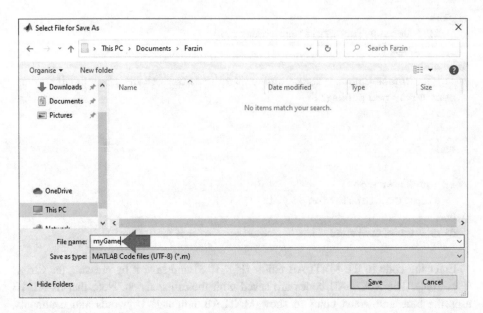

Fig. 10.4 Saving the entered code

Click the run button to run the program (Fig. 10.5). You can press the F5 key of your keyboard as well.

Fig. 10.5 press the run button
to run the entered code

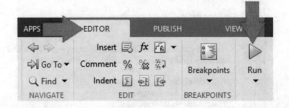

The message box shown in Fig. 10.6 may appear after running the code. If it is appeared, click the change folder or add to path button.

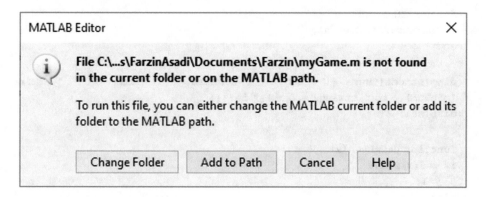

Fig. 10.6 MATLAB editor message

A sample run of the code is shown in Fig. 10.7.

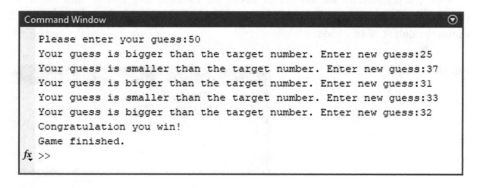

Fig. 10.7 Sample run of the entered code

For loop

The following program calculates the summation of even numbers from minNum to maxNum. This program uses a user defined function.

```
%file: EvenNumberSummation.m
%this program claculates the summation of
%even numbers from minNum to maxNum.

sum=0;
minNum=0;
maxNum=10;
```

```
for n=minNum:maxNum
   sum=sum+n*isEven(n);
end

disp(strcat("sum          of          numbers          from          ",string
(minNum)," to ",string(maxNum)," is:"))
disp(sum)

function y=isEven(x)
if mod(x,2)==0
   y=1;
else
   y=0;
end
end
```

Type the program in the MATLAB editor environment and save it with the name EvenNumberSummation.m. Then run the program by pressing the F5 key. Figure 10.8 shows the output of this code.

Fig. 10.8 Output of the code

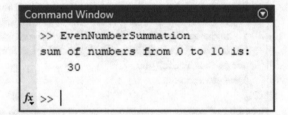

10.4   Switch-Case Control Statement

Let's study an example for switch-case control statement. The following code takes a number and shown the day of week which is associated with it. For instance, if the user enters 1, it shows Monday and if the user enters 4, it shows Thursday.

```
clc

day = input('Enter the number (1..7): ');

switch (day)
```

```
    case 1
       disp('Monday')
    case 2
       disp('Thuesday')
    case 3
       disp('Wednesday')
    case 4
       disp('Thursday')
    case 5
       disp('Friday')
    case 6
       disp('Saturday')
    case 7
       disp('Sunday')
    otherwise
       disp('Invalid number')
end
```

Write the code in MATLAB Editor (Fig. 10.9). Sample run of this code is shown in Fig. 10.10.

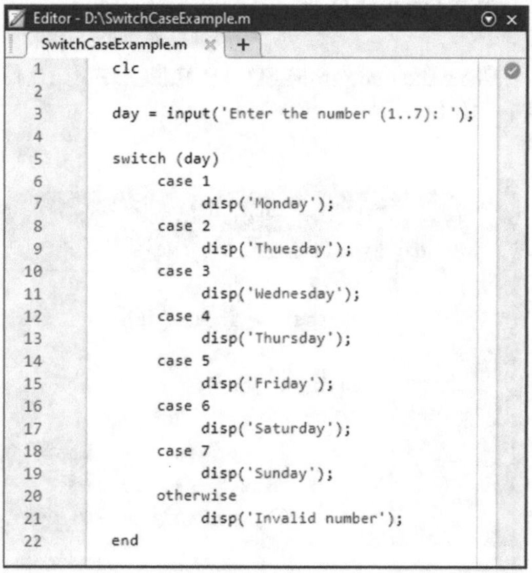

Fig. 10.9 Code is entered to MATLAB editor

Fig. 10.10 Sample run of
code in Fig. 10.9

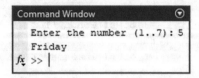

Let's study another example. Following code takes a number from the user and
determines whether it is less than 5, equal to 5 and bigger than 5.

```
clc
number = input('Enter an integer number (1..10)');

switch (number)
    case {1,2,3,4}
        result = 'Less than 5.';
    case {6,7,8,9,10}
        result = 'Bigger than 5.';
    otherwise
        result = 'Entered number is 5.';
end
disp(result);
```

Enter the code to MATLAB Editor (Fig. 10.11). Sample run of this code is shown in
Fig. 10.12.

```
Editor - D:\SwitchCaseExample2.m *                              ⊙ ✕
  SwitchCaseExample2.m *  ✕  +
  1       clc
  2       number = input('Enter an integer number (1..10)');
  3
  4       switch (number)
  5           case {1,2,3,4}
  6               result = 'Less than 5.';
  7           case {6,7,8,9,10}
  8               result = 'Bigger than 5.';
  9           otherwise
 10               result = 'Entered number is 5.';
 11       end
 12       disp(result);
```

Fig. 10.11 Code is entered to MATLAB editor

Fig. 10.12 Sample run of code in Fig. 10.11

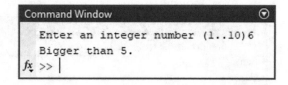

10.5   fprintf and disp Command

The code shown in Fig. 10.13 calculates the $\sum_{i=1}^{20} i^2$. Output of this code is shown in Fig. 10.14.

Fig. 10.13 Code to calculate the $\sum_{i=1}^{20} i^2$

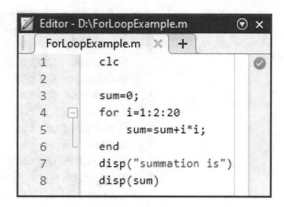

Fig. 10.14 Output of the code in Fig. 10.13

Let's change the above code in a way that it shows the partial summations ($\sum_{i=1}^{1} i^2, \sum_{i=1}^{2} i^2, \sum_{i=1}^{3} i^2, \ldots$) with suitable messages. Change the code in Fig. 10.13 to what shown in Fig. 10.15. Output of this code is shown in Fig. 10.16. fprintf is originally used to write data to text files. However, you can use it to display something on screen as well. %d is used to show signed integers numbers.

```
Editor - D:\ForLoopExample.m *                                        ⊙ ×
  ForLoopExample.m *   ✕   +
  1        clc
  2
  3        sum=0;
  4   ⊟    for i=1:2:20
  5            sum=sum+i*i;
  6            fprintf('Summation up to %d is %d\n',i,sum);
  7        end
```

Fig. 10.15 Code for calculation of partial summations

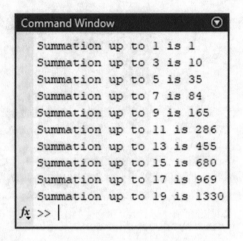

```
Command Window                        ⊙
     Summation up to 1 is 1
     Summation up to 3 is 10
     Summation up to 5 is 35
     Summation up to 7 is 84
     Summation up to 9 is 165
     Summation up to 11 is 286
     Summation up to 13 is 455
     Summation up to 15 is 680
     Summation up to 17 is 969
     Summation up to 19 is 1330
  fx >> |
```

Fig. 10.16 Output of the code in Fig. 10.15

You can obtain the result shown in Fig. 10.16 with the aid of the disp command as well (Fig. 10.17).

```
Editor - D:\ForLoopExample.m
ForLoopExample.m  +
1    clc
2
3    sum=0;
4    for i=1:2:20
5        sum=sum+i*i;
6        disp("Summation up to "+i+" is "+sum);
7    end
```

Fig. 10.17 Equivalent for code in Fig. 10.15

10.6 Functions

In this section you learn how to write functions. Assume that we need a function to calculate $x^3 + 3x^2 + 5$. The code shown in Fig. 10.18 do this job for us. Name of the function in Fig. 10.18 is myFunction. The name that is used to save the code must be the same as the function name. For instance, the code shown in Fig. 10.18 must be saved with the name myFunction.m on your hard disk. Sample run of this code is shown in Fig. 10.19.

Fig. 10.18 Function to calculate $f(x) = x^3 + 3x^2 + 5$

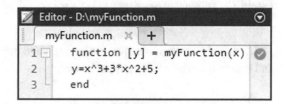

```
Editor - D:\myFunction.m
myFunction.m  +
1    function [y] = myFunction(x)
2    y=x^3+3*x^2+5;
3    end
```

Fig. 10.19 Sample run of the function in Fig. 10.18

Let's study another example. The code shown in Fig. 10.20 defines a function to calculate $z = \begin{bmatrix} z_1 \\ z_2 \end{bmatrix} = \begin{bmatrix} x_1^3 + 3x_2^2 + 5 \\ \sin(x_1) + x_2^2 \end{bmatrix}$. Sample run of this function is shown in Fig. 10.21.

```
Editor - D:\myFunction2.m

myFunction2.m   ☒   +

1    function [z] = myFunction2(x)
2        z(1) =x(1)^3+3*x(2)^2+5;
3        z(2)=sin(x(1))+x(2)^2;
4    end
```

Fig. 10.20 Function to calculate $z = \begin{bmatrix} z_1 \\ z_2 \end{bmatrix} = \begin{bmatrix} x_1^3+3x_2^2+5 \\ \sin(x_1)+x_2^2 \end{bmatrix}$

Let's study another example. Following code defines a function that returns the mean and standard deviation of an input vector.

```
%calculates the mean and standard deviation of a vector
function [m,s] = stat(x)
   n = length(x);
   m = sum(x)/n;
   s = sqrt(sum((x-m).^2/n));
end
```

Command Window

```
>> myFunction2([1 2])

ans =

    18.0000    4.8415

>> myFunction2([1;2])

ans =

    18.0000    4.8415

fx >> |
```

Fig. 10.21 Sample run of function in Fig. 10.20

Enter the code into the MATLAB Editor environment (Fig. 10.22) and save it with the name stat.m. Note that this code starts with a comment line.

Editor - D:\stat.m

stat.m

```
1    %calculates the mean and standard deviation of a vector
2    function [m,s] = stat(x)
3        n = length(x);
4        m = sum(x)/n;
5        s = sqrt(sum((x-m).^2/n));
6    end
```

Fig. 10.22 Function to calculate the mean and standard deviation

Now return to the MATLAB environment and enter the commands shown in Fig. 10.23. Note that help stat in Fig. 10.23 returned the first comment line.

```
Command Window                                                    ⊙

  >> help stat
   calculates the mean and standard deviation of a vector

  >> [Mean,StdDev]=stat([1 3 5 9 9 4])

  Mean =

      5.1667


  StdDev =

      2.9674

fx >> |
```

Fig. 10.23 Sample run of function in Fig. 10.22

10.7   Calculation of Fourier Series Coefficients

Let's write a MATLAB code to calculate the Fourier series coefficients of the triangular waveform shown in Fig. 10.24. This helps you to improve your MATLAB programming.

Fig. 10.24 Triangular waveform

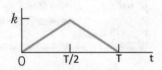

The function shown in Fig. 10.24 can be written as:

$$f(x) = \begin{cases} \frac{2k}{L}x & if\ 0 < x < \frac{L}{2} \\ \frac{2k}{L}(L-x) & if\ \frac{L}{2} < x < L \end{cases} \tag{10.1}$$

Assume that $T = 2\pi$ and $k = 1$. Following code calculates the Fourier series coefficients up to the 10th harmonic.

```
clc
clear all

T=2*pi;
w0=2*pi/T;
k=1;
N=10;     %number of terms

syms t
f1=2*k/T*t;
f2=2*k/T*(T-t);

a0=eval(1/T*(int(f1,0,T/2)+int(f2,T/2,T)));

a=zeros(1,N);
for n=1:N
    a(n)=2/T*(int(f1*cos(n*w0*t),0,T/2)+int(f2*cos(n*w0*t),T/2,T));
end

b=zeros(1,N);
for n=1:N
    b(n)=2/T*(int(f1*sin(n*w0*t),0,T/2)+int(f2*sin(n*w0*t),T/2,T));
end

disp("Average (DC component)= "+a0);
h=zeros(1,N);
for i=1:N
    h(i)=sqrt(a(i)^2+b(i)^2);
    disp("Value of "+i+"th harmonic is="+h(i));
end
```

Enter the code to MATLAB editor (Fig. 10.25). Output of the code is shown in Fig. 10.26.

Fig. 10.25   Code is entered to the MATLAB editor

Fig. 10.26   Output of the code
in Fig. 10.25

10.8   Newton–Raphson Method

In this section we want to write a code for Newton–Raphson method. Following code uses the Newton–Raphson method in order to find the root of $f(x) = 2^x - 5x + 2$. This code stops either after 20 iterations or when the difference between two consecutive guess is less than 0.0001.

```
clc
clear all

f=@(x) 2^x-5*x+2;
df=@(x) log(2)*(2^x)-5;

n=20;   % Number of iterations
e=1e-4; % Acceptable error
x0=0;   % Inital point

if df(x0)~=0
   for i=1:n
      x1=x0-f(x0)/df(x0);
      fprintf('x%d=%.6f\n',i,x1);
      if (x1-x0)<e
         break
      end
      x0=x1;
   end
else
   disp('Newton Raphson failed...');
end
```

Enter the code into the MATLAB Editor (Fig. 10.27) and run it. Output of this code is shown in Fig. 10.28. The code stops after 4 iterations.

```
Editor - D:\NewtonRaphsonMethod.m                              ⊙

  NewtonRaphsonMethod.m  ✕  +

   1         clc
   2         clear all
   3
   4         f=@(x) 2^x-5*x+2;
   5         df=@(x) log(2)*(2^x)-5;
   6
   7         n=20;    % Number of iterations
   8         e=1e-4;  % Acceptable error
   9         x0=0;    % Inital point
  10
  11         if df(x0)~=0
  12    ⊟       for i=1:n
  13               x1=x0-f(x0)/df(x0);
  14               fprintf('x%d=%.6f\n',i,x1);
  15               if (x1-x0)<e
  16                   break
  17               end
  18               x0=x1;
  19           end
  20         else
  21           disp('Newton Raphson failed...');
  22         end
```

Fig. 10.27 Code is entered to MATLAB editor

Fig. 10.28 Output of the code
in Fig. 10.27

```
Command Window  ⊙

    x1=0.696564
    x2=0.732115
    x3=0.732244
    x4=0.732244
fx >> |
```

10.9   Bisection Method

In this section we want to write a code for bisection method. Following code uses the bisection method in order to find the root of $f(x) = e^x - 5x + 2$. This code stops either after 30 iterations or when the difference between two consecutive guess is less than 0.0001.

```
%Bisection Code
clc
clear all

f=@(x) exp(x)-5*x+2; % function that we want to find its roots
a=0;              % search in the [a,b] interval
b=1;
n=30;             % number of iterarions
e=0.0001;         % error

if f(a)*f(b)<0
    for i=1:n
        c=(a+b)/2;
        fprintf('P%d=%.4f\n',i,c)
        if abs(c-b)<e||abs(c-a)<e
            break
        end
        if f(a)*f(c)<0
            b=c;
        elseif f(b)*f(c)<0
            a=c;
        end
    end
end
```

Enter the code into the MATLAB Editor (Fig. 10.29) and run it. Output of this code is shown in Fig. 10.30.

```
Editor - D:\BiSection.m                                                    ⊙ ×
BiSection.m   ✕   +
1        %Bisection Code
2        clc
3        clear all
4
5        f=@(x) exp(x)-5*x+2; %function that we want to find its roots
6        a=0;                 % search in the [a,b] interval
7        b=1;
8        n=30;                % number of iterarions
9        e=0.0001;            % error
10
11       if f(a)*f(b)<0
12           for i=1:n
13               c=(a+b)/2;
14               fprintf('P%d=%.4f\n',i,c)
15               if abs(c-b)<e||abs(c-a)<e
16                   break
17               end
18               if f(a)*f(c)<0
19                       b=c;
20               elseif f(b)*f(c)<0
21                       a=c;
22               end
23           end
24       end
```

Fig. 10.29 Code is entered to MATLAB editor

Fig. 10.30 Output of code in
Fig. 10.29

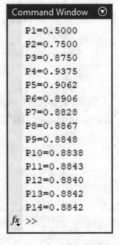

```
Command Window   ⊙
    P1=0.5000
    P2=0.7500
    P3=0.8750
    P4=0.9375
    P5=0.9062
    P6=0.8906
    P7=0.8828
    P8=0.8867
    P9=0.8848
    P10=0.8838
    P11=0.8843
    P12=0.8840
    P13=0.8842
    P14=0.8842
fx >>
```

10.10 Taking the Equation of the Function from the User

In the previous examples we entered the function into the code. You can take the function from the user in run time as well. In this example you will learn how to take the function from the user in run time. Change the code of previous example to what shown below:

```
%Bisection Code
clc
clear all

str=input('please enter f(x)=',"s");
f=str2func("@(x) "+str);   % function that we want to find its roots
a=0;                       % search in the [a,b] interval
b=1;
n=30;                      % number of iterarions
e=0.0001;                   % error

if f(a)*f(b)<0
    for i=1:n
        c=(a+b)/2;
        fprintf('P%d=%.4f\n',i,c)
        if abs(c-b)<e||abs(c-a)<e
            break
        end
        if f(a)*f(c)<0
            b=c;
        elseif f(b)*f(c)<0
            a=c;
        end
    end
end
```

The function str2func convert the entered string into a mathematical function. Enter the code into the MATLAB Editor (Fig. 10.31). Sample output of this code is shown in Fig. 10.32.

```
Editor - D:\BiSection.m
  BiSection.m   ✕   +
1        %Bisection Code
2        clc
3        clear all
4
5        str=input('please enter f(x)=',"s");
6        f=str2func("@(x) "+str);   % function that we want to find its roots
7        a=0;                       % search in the [a,b] interval
8        b=1;
9        n=30;                      % number of iterarions
10       e=0.0001;                  % error
11
12       if f(a)*f(b)<0
13           for i=1:n
14               c=(a+b)/2;
15               fprintf('P%d=%.4f\n',i,c)
16               if abs(c-b)<e||abs(c-a)<e
17                   break
18               end
19               if f(a)*f(c)<0
20                       b=c;
21               elseif f(b)*f(c)<0
22                       a=c;
23               end
24           end
25       end
```

Fig. 10.31   Code is entered to MATLAB editor

Fig. 10.32   Sample run of
code in Fig. 10.31

```
Command Window
  please enter f(x)=x^2-2.7*x+1.4
  P1=0.5000
  P2=0.7500
  P3=0.6250
  P4=0.6875
  P5=0.7188
  P6=0.7031
  P7=0.6953
  P8=0.6992
  P9=0.7012
  P10=0.7002
  P11=0.6997
  P12=0.7000
  P13=0.7001
  P14=0.7000
fx >> |
```

10.11 Range-Kutta 45

In this section we want to use the Range-Kutta (4, 5) method to solve the following differential equation for [0, 1] time interval:

$$\ddot{y} + 5\dot{y} + 6y = \cos(t),\ y(0) = 3\dot{y}(0) = -1 \qquad (10.2)$$

Let's convert the given equation into a state space equation:

$$\begin{cases} y_1 = y \\ y_2 = \dot{y} \end{cases} \qquad (10.3)$$

$$\begin{cases} \dot{y}_1 = y_2 \\ \dot{y}_2 = -6y_1 - 5y_2 + \cos(t) \end{cases},\ y_1(0) = 3,\ y_2(0) = -1 \qquad (10.4)$$

Let's enter the obtained function into the MATLAB environment (Fig. 10.33).

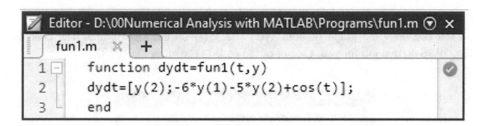

```
Editor - D:\00Numerical Analysis with MATLAB\Programs\fun1.m

fun1.m    +

1   function dydt=fun1(t,y)
2   dydt=[y(2);-6*y(1)-5*y(2)+cos(t)];
3   end
```

Fig. 10.33 State space equations are entered to MATLAB

The code shown Fig. 10.34 solves the differential equation with given initial conditions for [0,1] time interval. Output of this code is shown in Fig. 10.35. Figure 10.35 show the graph of $y_1(t)$.

```
Command Window

>> [t,y] = ode45(@fun1,[0 1],[3;-1]);
>> plot(t,y(:,1),'-x')
>> xlabel('Time t')
>> ylabel('Solution y')
fx >>
```

Fig. 10.34 Solving the state space equations with ode45 and plotting the $y_1(t)$

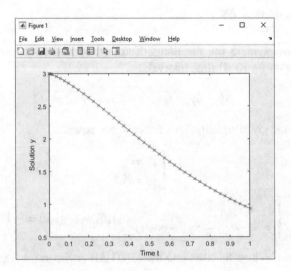

Fig. 10.35 Output of code in Fig. 10.34

You can draw the graph of both $y_1(t)$ and $y_2(t)$ on the same graph. The code in Fig. 10.36 do this job for you. Output of the code is shown in Fig. 10.37.

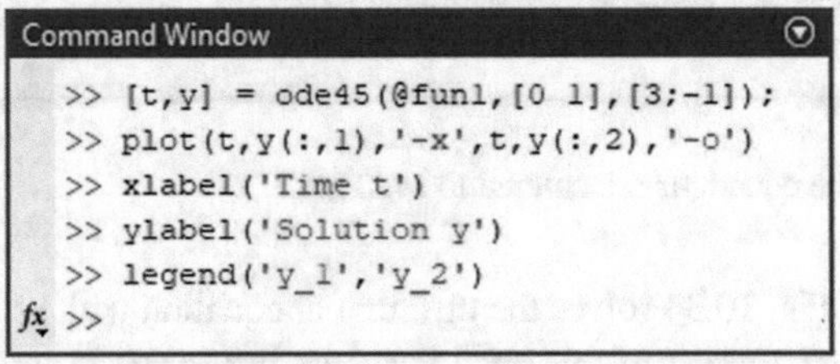

```
Command Window                                    ⊙
 >> [t,y] = ode45(@fun1,[0 1],[3;-1]);
 >> plot(t,y(:,1),'-x',t,y(:,2),'-o')
 >> xlabel('Time t')
 >> ylabel('Solution y')
 >> legend('y_1','y_2')
fx >>
```

Fig. 10.36 Solving the state space equations with ode45 and plotting the $y_1(t)$ and $y_2(t)$

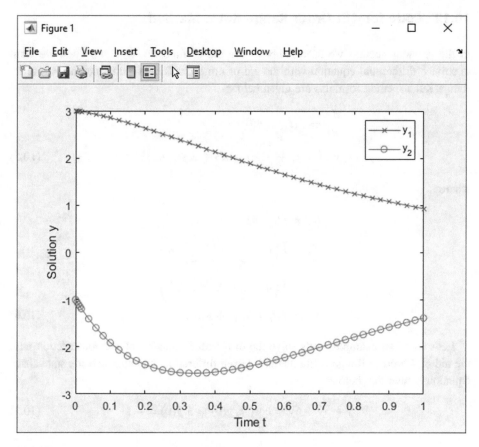

Fig. 10.37 Output of the code in Fig. 10.36

10.12 Code for 4th Order Range-Kutta Method

In the previous section we used ready to use ode45 function. In this section we want to solve a differential equation with the aid of 4th order Range-Kutta method. 4th order Range-Kutta method formulas are given below:

$$\frac{dy}{dt} = f(t, y),\ y(0) = y_0$$

$$y_{i+1} = y_i + \frac{1}{6}(K_1 + K_2 + K_3 + K_4)h \tag{10.5}$$

where

$$K_1 = f(t_i, y_i)$$

$$K_2 = f\left(t_i + \frac{1}{2}h,\ y_i + \frac{1}{2}K_1h\right)$$

$$K_3 = f\left(t_i + \frac{1}{2}h,\ y_i + \frac{1}{2}K_2h\right)$$

$$K_4 = f(t_i + h,\ y_i + K_3h) \tag{10.6}$$

Let's study an example. Let's solve the differential equation of previous section with the aid of 4th order Range-Kutta method. Given differential equation and it's state space equation is rewritten below:

$$\ddot{y} + 5\dot{y} + 6y = \cos(t),\ y(0) = 3\ \dot{y}(0) = -1 \tag{10.2}$$

$$\begin{cases} y_1 = y \\ y_2 = \dot{y} \end{cases} \tag{10.3}$$

$$\begin{cases} \dot{y}_1 = y_2 \\ \dot{y}_2 = -6y_1 - 5y_2 + \cos(t) \end{cases},\ y_1(0) = 3,\ y_2(0) = -1 \tag{10.4}$$

Following code solves this equation with the aid of 4th order Range-Kutta method.

```
clc
clear all
format shortG

h=0.025;   % Time step
T=0:h:1;   %solution is calculated for [0,1]
y0=[3;-1]; %initial point

N=length(T);
y=zeros(2,N);
y(:,1)=y0;

for n=1:N-1
   K1=fun2(y(:,n), T(n));
   K2=fun2(y(:,n)+h/2*K1,T(n)+h/2);
   K3=fun2(y(:,n)+h/2*K2,T(n)+h/2);
   K4=fun2(y(:,n)+h*K3,T(n)+h);
   y(:,n+1)=y(:,n)+h/6*(K1+2*K2+2*K3+K4);
end

plot(T,y(1,:),'b',T,y(2,:),'r')
legend('y1=y','y2=dy/dt')

function dydt=fun2(y,t)
   dydt=[y(2);-6*y(1)-5*y(2)+cos(t)];
end
```

Enter the code into the MATLAB Editor (Fig. 10.38) and run the code. Output of the code is shown in Fig. 10.39. The obtained result is quite similar to Fig. 10.37.

```
Editor - D:\00Numerical Analysis with MATLAB\Programs\RK4.m        ⊙  ×

  RK4.m    ×   +
  1         clc
  2         clear all
  3         format shortG
  4
  5         h=0.025;      % Time step
  6         T=0:h:1;      %solution is calculated for [0,1]
  7         y0=[3;-1];    %initial point
  8
  9         N=length(T);
 10         y=zeros(2,N);
 11         y(:,1)=y0;
 12
 13    ⊟    for n=1:N-1
 14            K1=fun2(y(:,n), T(n));
 15            K2=fun2(y(:,n)+h/2*K1,T(n)+h/2);
 16            K3=fun2(y(:,n)+h/2*K2,T(n)+h/2);
 17            K4=fun2(y(:,n)+h*K3,T(n)+h);
 18            y(:,n+1)=y(:,n)+h/6*(K1+2*K2+2*K3+K4);
 19         end
 20
 21         plot(T,y(1,:),'b',T,y(2,:),'r')
 22         legend('y1=y','y2=dy/dt')
 23
 24    ⊟    function dydt=fun2(y,t)
 25            dydt=[y(2);-6*y(1)-5*y(2)+cos(t)];
 26         end
 27
```

Fig. 10.38 Code is entered to MATLAB editor

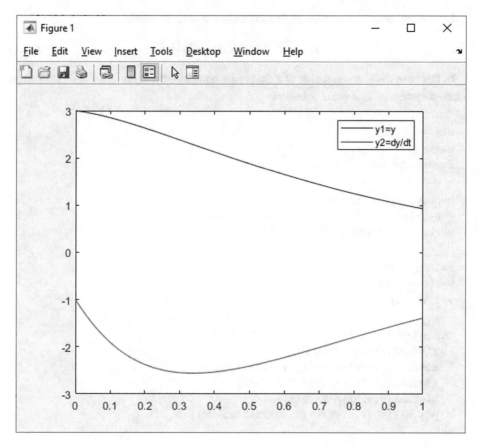

Fig. 10.39 Output of code in Fig. 10.38

Let's study another example. In this example we want to solve:

$$\dot{y} = -y^3 + y\sin(2t),\ y(0) = 1 \qquad (10.7)$$

In this case the given equation is a first order equation. So, there is no need to convert it into a system of first order equation.

```
clc
clear all
format shortG


h=0.05;     % Time step
T=0:h:10;   %solution is calculated for [0,1]
y0=1;       %initial point

N=length(T);
y=zeros(1,N);
y(:,1)=y0;


for n=1:N-1
   K1=fun3(y(:,n), T(n));
   K2=fun3(y(:,n)+h/2*K1,T(n)+h/2);
   K3=fun3(y(:,n)+h/2*K2,T(n)+h/2);
   K4=fun3(y(:,n)+h*K3,T(n)+h);
   y(:,n+1)=y(:,n)+h/6*(K1+2*K2+2*K3+K4);
end


plot(T,y(1,:),'b');


function dydt=fun3(y,t)
   dydt=-y^3+y*sin(2*t);
end
```

Enter the code into the MATLAB Editor (Fig. 10.40) and run the code. Output of the code is shown in Fig. 10.41.

Fig. 10.40 Code is entered to
MATLAB editor

```
Editor - D:\00Numerical Analysis with MATLAB\Programs\RK4.m
RK4.m    +
1    clc
2    clear all
3    format shortG
4
5    h=0.05;          % Time step
6    T=0:h:10;       %solution is calculated for [0,1]
7    y0=1;           %initial point
8
9    N=length(T);
10   y=zeros(1,N);
11   y(:,1)=y0;
12
13   for n=1:N-1
14       K1=fun3(y(:,n), T(n));
15       K2=fun3(y(:,n)+h/2*K1,T(n)+h/2);
16       K3=fun3(y(:,n)+h/2*K2,T(n)+h/2);
17       K4=fun3(y(:,n)+h*K3,T(n)+h);
18       y(:,n+1)=y(:,n)+h/6*(K1+2*K2+2*K3+K4);
19   end
20
21   plot(T,y(1,:),'b');
22
23   function dydt=fun3(y,t)
24       dydt=-y^3+y*sin(2*t);
25   end
26
```

Fig. 10.41 Output of code in
Fig. 10.40

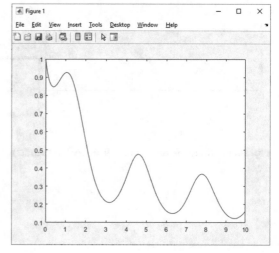

Let's check the obtained result with the aid of Simulink. Simulink model of $\dot{y} = -y^3 + y\sin(2t)$ is shown in Fig. 10.42. Settings of integrator and sine wave blocks are shown in Figs. 10.43 and 10.44.

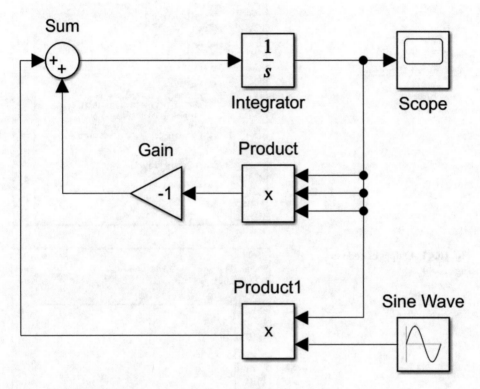

Fig. 10.42   Simulink diagram for $\dot{y} = -y^3 + y\sin(2t)$

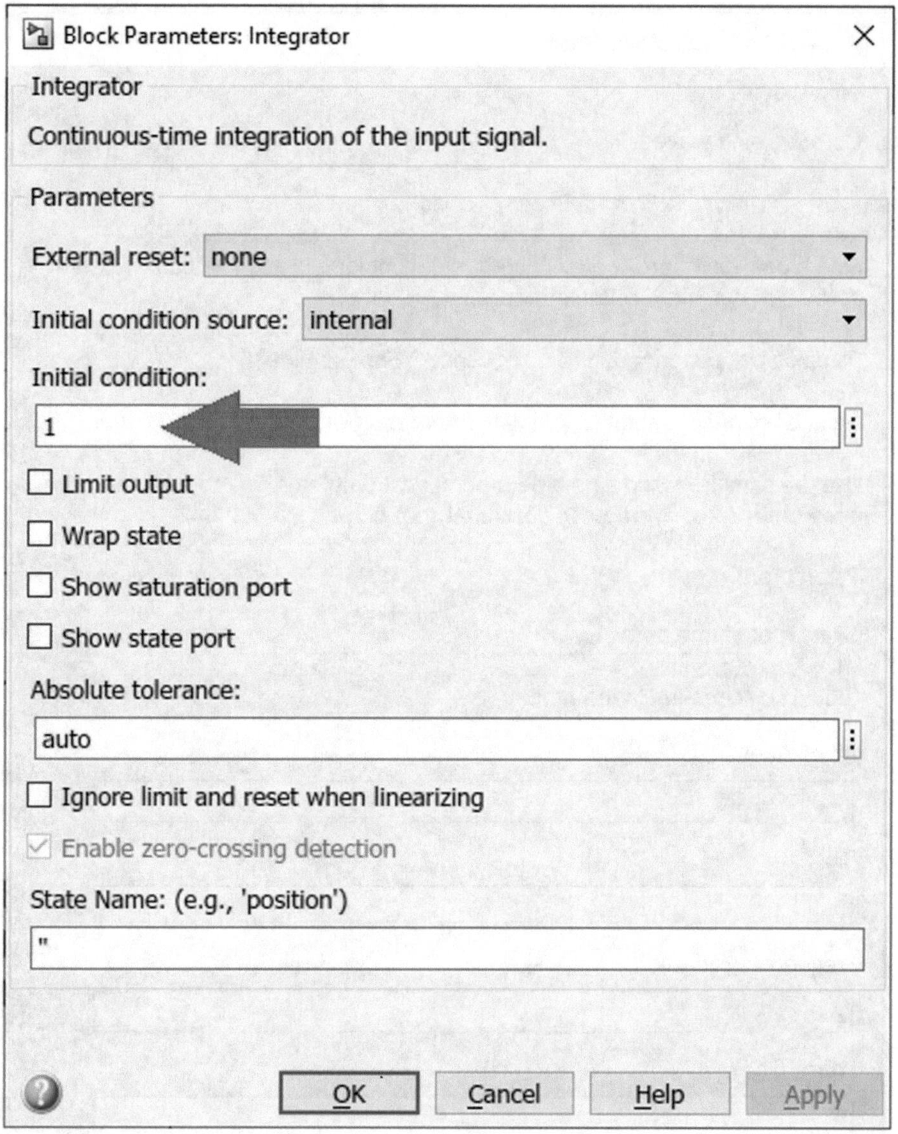

Fig. 10.43   Settings of integrator block in Fig. 10.42

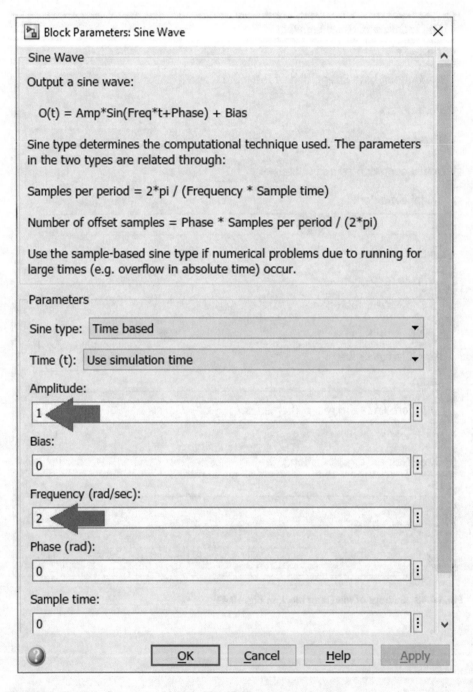

Fig. 10.44   Settings of sine wave block in Fig. 10.42

Press the Ctrl+E and change the settings to what shown in Fig. 10.45. Then run the simulation. Output is shown in Fig. 10.46. You can use the cursors in order to read different points of Figs. 10.41 and 10.46 and ensure that they are the same.

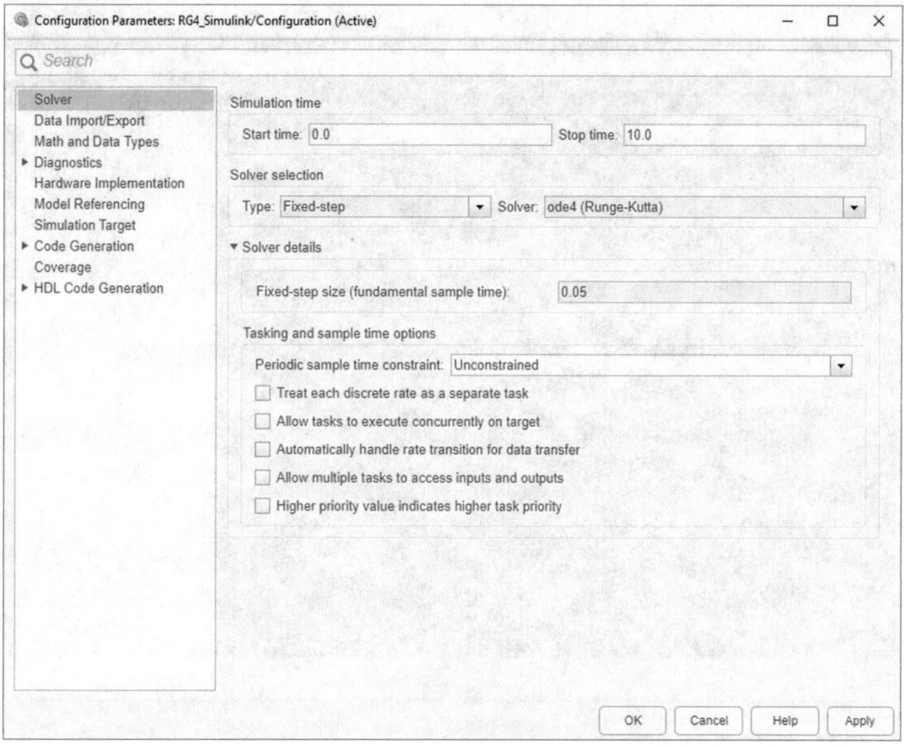

Fig. 10.45 Solver settings

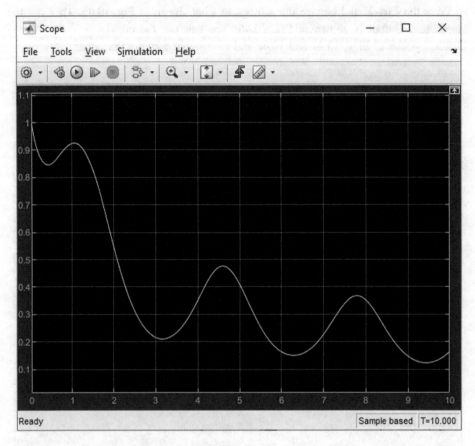

Fig. 10.46   Simulation result

Exercises

1. Use hand analysis to calculate the Fourier series coefficients of the waveform shown in Fig. 10.24.
2. Write a MATLAB code to solve the $x^2 - x - 2 = 0$ using the bisection method. Test your code using the initial points ($\times 0 = 1.5, \times 1 = 2.5$) and ($\times 0 = -1.5, \times 1 = --0.5$).
3. Write a MATLAB code to solve the $x^3 - 6x^2 + 11x - 6 = 0$ using the Newton–Raphson method. Test your code using the initial point $\times 0 = 6$.
4. Write a MATLAB code to solve the $\dot{y} = -y^3 + y\sin(2t)$, $y(0) = 1$ using the 3rd order Range-Kutta method.

References for Further Study

1. Chapman, S.: MATLAB Programming for Engineers, 6th edition, Cengage, 2019.
2. Hahn, B., Valentine, D.: Essential MATLAB for Engineers and Scientists, 7th edition, Academic Press, 2019.
3. Moore, H.: MATLAB for Engineers, 5th edition, Pearson, 2017.

Optimization with MATLAB® 11

11.1 Introduction

Mathematical optimization (alternatively, optimization or mathematical programming) is the selection of a best element (with regard to some criteria) from some set of available alternatives. In the simplest case, an optimization problem consists of maximizing or minimizing a real function by systematically choosing input values from within an allowed set and computing the value of the function.

MATLAB's Optimization Toolbox™ consist of many ready to use functions to solve optimization problems. In this chapter you will learn how to solve basic optimization problems with MATLAB. Note that MATLAB solves minimization problems. If you want to maximize something, you need to convert it into a minimization problem. Fortunately this can be done easily: $\dfrac{\max f(x)}{x} = \dfrac{\min -f(x)}{x}$. So, when you want to solve a maximization problem with MATLAB, you need to enter the negative of the given cost function.

11.2 Local Minimum

fmincon command finds a constrained minimum of a function of several variables. fmincon attempts to solve problems of the form:

$$\min_{x} f(x) \quad \text{subject to:} \begin{cases} A \times x \leq B, \quad A_{eq} \times x = B_{eq} \ (linear\ constraints) \\ C(x) \leq 0, \quad C_{eq}(x) = 0 \quad (nonlinear\ constraints) \ . \\ LB \leq x \leq UB \quad\quad\quad (bounds) \end{cases} \quad (11.1)$$

© The Author(s), under exclusive license to Springer Nature Switzerland AG 2023
F. Asadi, *Applied Numerical Analysis with MATLAB®/Simulink®*,
Synthesis Lectures on Engineering, Science, and Technology,
https://doi.org/10.1007/978-3-031-19366-8_11

Let's study an example. Assume $f(x) = x^3 + 30 \sin(x^2) + 10\cos(x)$. Let's draw the graph of this function for $[0, \pi]$ interval. The commands shown in Fig. 11.1 do this job for us. Output of this code is shown in Fig. 11.2.

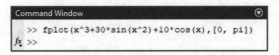

Fig. 11.1 Drawing the graph of $f(x) = x^3 + 30 \sin\left(x^2\right) + 10\cos(x)$ for $[0, \pi]$ interval

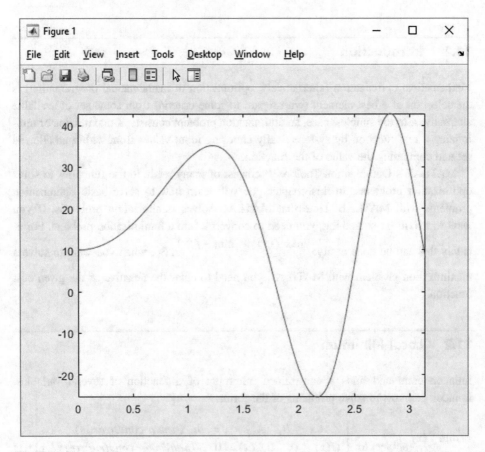

Fig. 11.2 Output of the code in Fig. 11.1

You can click on the optimum points of the graph in order to see their values (Fig. 11.3).

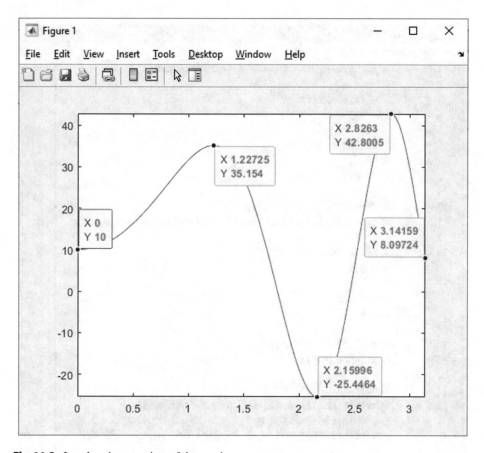

Fig. 11.3 Local optimum points of the graph

The following code starts with initial point x0 = 1 and tries to find the local minimum. Enter the code into MATLAB editor (Fig. 11.4). The only constraint that we have in this problem is $0 < x < \pi$.

```
clc
objective = @(x) x^3+30*sin(x^2)+10*cos(x);
% initial guess
x0 = 1;
% variable bounds
lb = 0;
ub = pi;
% show initial objective
disp(['Initial Objective: ' num2str(objective(x0))])
% linear constraints
```

```
Editor - D:\OptExample.m *                                            ⊙  ×
   OptExample.m *   ×   +
 1          objective = @(x) x^3+30*sin(x^2)+10*cos(x);                  ⊘
 2
 3          % initial guess
 4          x0 = 1;
 5
 6          % variable bounds
 7          lb = 0;
 8          ub = pi;
 9
10          % show initial objective
11          disp(['Initial Objective: ' num2str(objective(x0))])
12
13          % linear constraints
14          A = [];
15          b = [];
16          Aeq = [];
17          beq = [];
18
19          % nonlinear constraints
20          nonlincon = [];
21
22          % optimize with fmincon
23          %[X,FVAL,EXITFLAG,OUTPUT,LAMBDA,GRAD,HESSIAN]
24          % = fmincon(FUN,X0,A,B,Aeq,Beq,LB,UB,NONLCON,OPTIONS)
25          x = fmincon(objective,x0,A,b,Aeq,beq,lb,ub,nonlincon);
26
27          % show final objective
28          disp(['Final Objective: ' num2str(objective(x))])
29
30          % print solution
31          disp('Solution')
32          disp(['x = ' num2str(x)])
```

Fig. 11.4 Code for calculation of minimum point of $f(x)$ with initial point of $x0 = 1$

```
A = [];
b = [];
Aeq = [];
beq = [];
% nonlinear constraints
nonlincon = [];
% optimize with fmincon
```

```
%[X,FVAL,EXITFLAG,OUTPUT,LAMBDA,GRAD,HESSIAN]
% = fmincon(FUN,X0,A,B,Aeq,Beq,LB,UB,NONLCON,OPTIONS)
x = fmincon(objective,x0,A,b,Aeq,beq,lb,ub,nonlincon);
% show final objective
disp(['Final Objective: ' num2str(objective(x))])
% print solution
disp('Solution')
disp(['x = ' num2str(x)])
```

Now run the code. Output of the code is shown in Fig. 11.5. Starting with the initial point of x0 = 1, MATLAB found x = 0.00014114 as the argument which minimize our cost function $f(x) = x^3 + 30\sin(x^2) + 10\cos(x)$. According to Fig. 11.5, value of cost function at x = 0.00014114 is equal to 10.

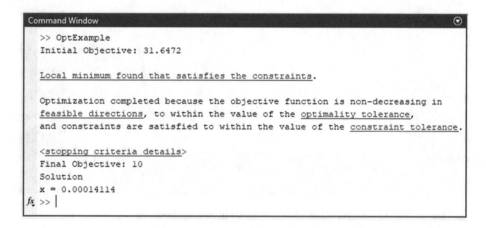

Fig. 11.5 Output of the code in Fig. 11.4

Starting point x0 and found solution are shown in Fig. 11.6. As shown in Fig. 11.6, the starting point x0 = 1 is between two optimum points: x = 0 with value of 10 and x = 1.21897 with value of 35.1493. In this case MATLAB returns x = 0 because cost function has a smaller value there.

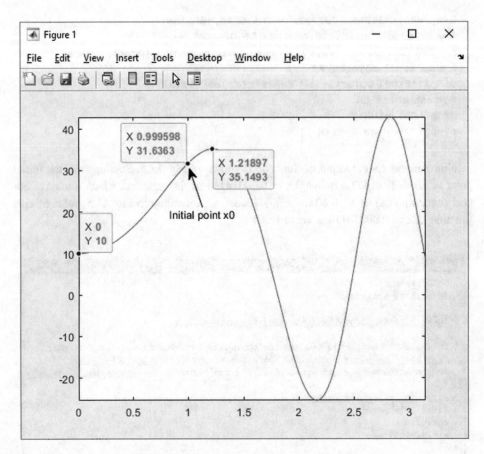

Fig. 11.6 Selected initial point and optimum points in its neighborhood

Now change the initial point to 2 (Fig. 11.7).

```
Editor - D:\OptExample.m                                    ⊙  ✕
  OptExample.m  ✕  +
1       objective = @(x) x^3+30*sin(x^2)+10*cos(x);              ✓
2
3       % initial guess
4       x0 = 2;|
5
6       % variable bounds
7       lb = 0;
8       ub = pi;
9
10      % show initial objective
11      disp(['Initial Objective: ' num2str(objective(x0))])
12
13      % linear constraints
14      A = [];
15      b = [];
16      Aeq = [];
17      beq = [];
18
19      % nonlinear constraints
20      nonlincon = [];
21
22  ⊟   % optimize with fmincon
23      %[X,FVAL,EXITFLAG,OUTPUT,LAMBDA,GRAD,HESSIAN]
24      % = fmincon(FUN,X0,A,B,Aeq,Beq,LB,UB,NONLCON,OPTIONS)
25      x = fmincon(objective,x0,A,b,Aeq,beq,lb,ub,nonlincon);
26
27      % show final objective
28      disp(['Final Objective: ' num2str(objective(x))])
29
30      % print solution
31      disp('Solution')
32      disp(['x = ' num2str(x)])
```

Fig. 11.7 Code for calculation of minimum point of $f(x)$ with initial point of x0 = 2

Run the code. Output of the code is shown in Fig. 11.8. Starting with initial point of x0 = 2, MATLAB found x = 2.1607 as the argument which minimize our cost function $f(x) = x^3 + 30\sin(x^2) + 10\cos(x)$. According to Fig. 11.8, value of cost function at x = 2.1607 is equal to −25.4466.

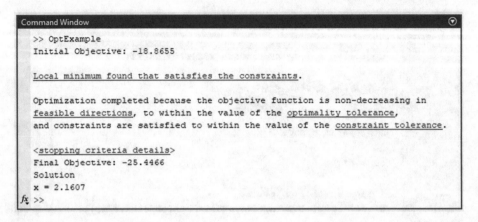

Fig. 11.8 Output of the code in Fig. 11.7

Starting point x0 and found solution are shown in Fig. 11.9. As shown in Fig. 11.9, the starting point x0 = 2 is between two optimum points: x = 1.21897 with value of 35.1493 and x = 2.16221 with value of −25.4459. In this case MATLAB returns x = 2.16221 because cost function has a smaller value there.

Fig. 11.9 Selected initial point and optimum points in its neighborhood

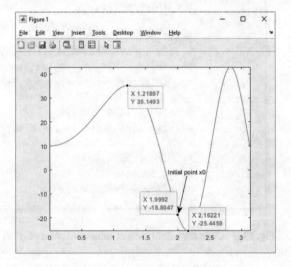

Change the initial point to 2.5 (Fig. 11.10).

```
    Editor - D:\00Numerical Analysis with MATLAB\Programs\OptExample.m

    OptExample.m  ×  +

 1      objective = @(x) x^3+30*sin(x^2)+10*cos(x);
 2
 3      % initial guess
 4      x0 = 2.5;
 5
 6      % variable bounds
 7      lb = 0;
 8      ub = pi;
 9
10      % show initial objective
11      disp(['Initial Objective: ' num2str(objective(x0))])
12
13      % linear constraints
14      A = [];
15      b = [];
16      Aeq = [];
17      beq = [];
18
19      % nonlinear constraints
20      nonlincon = [];
21
22      % optimize with fmincon
23      %[X,FVAL,EXITFLAG,OUTPUT,LAMBDA,GRAD,HESSIAN]
24      % = fmincon(FUN,X0,A,B,Aeq,Beq,LB,UB,NONLCON,OPTIONS)
25      x = fmincon(objective,x0,A,b,Aeq,beq,lb,ub,nonlincon);
26
27      % show final objective
28      disp(['Final Objective: ' num2str(objective(x))])
29
30      % print solution
31      disp('Solution')
32      disp(['x = ' num2str(x)])
```

Fig. 11.10 Code for calculation of minimum point of $f(x)$ with initial point of x0 = 2.5

Run the code. Output of the code is shown in Fig. 11.11. Starting with initial point of x0 = 2.5, MATLAB found x = 2.1607 as the argument which minimize our cost function $f(x) = x^3 + 30\sin(x^2) + 10\cos(x)$. According to Fig. 11.11, value of cost function at x = 2.1607 is equal to −25.4466.

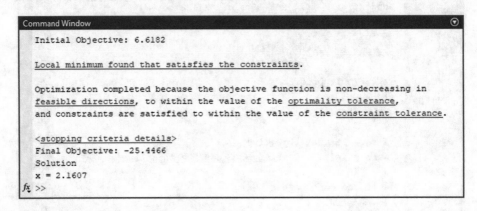

Fig. 11.11 Output of the code in Fig. 11.10

Starting point x0 and found solution are shown in Fig. 11.12. As shown in Fig. 11.12, the starting point x0 = 2.5 is between two optimum points: x = 2.16221 with value of −25.4459 and x = 2.8369 with value of 42.7291. In this case MATLAB returns x = 2.16221 because cost function has a smaller value there.

Fig. 11.12 Selected initial point and optimum points in its neighborhood

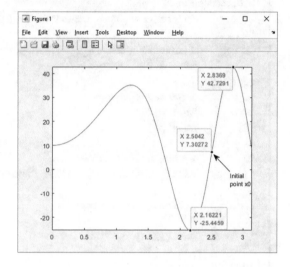

Change the initial point to 2.9 (Fig. 11.13).

```
Editor - D:\OptExample.m                                              ⊙ ✕
  OptExample.m  ✕   +
   1        objective = @(x) x^3+30*sin(x^2)+10*cos(x);              ✓
   2
   3        % initial guess
   4        x0 = 2.9;|        ⬅
   5
   6        % variable bounds
   7        lb = 0;
   8        ub = pi;
   9
  10        % show initial objective
  11        disp(['Initial Objective: ' num2str(objective(x0))])
  12
  13        % linear constraints
  14        A = [];
  15        b = [];
  16        Aeq = [];
  17        beq = [];
  18
  19        % nonlinear constraints
  20        nonlincon = [];
  21
  22   ⊟    % optimize with fmincon
  23        %[X,FVAL,EXITFLAG,OUTPUT,LAMBDA,GRAD,HESSIAN]
  24        % = fmincon(FUN,X0,A,B,Aeq,Beq,LB,UB,NONLCON,OPTIONS)
  25        x = fmincon(objective,x0,A,b,Aeq,beq,lb,ub,nonlincon);
  26
  27        % show final objective
  28        disp(['Final Objective: ' num2str(objective(x))])
  29
  30        % print solution
  31        disp('Solution')
  32        disp(['x = ' num2str(x)])
```

Fig. 11.13   Code for calculation of minimum point of $f(x)$ with initial point of $x0 = 2.9$

Run the code. Output of the code is shown in Fig. 11.14. Starting with initial point of $x0 = 2.9$, MATLAB found $x = 3.1416$ as the argument which minimize our cost function $f(x) = x^3 + 30 \sin(x^2) + 10\cos(x)$. According to Fig. 11.15, value of cost function at $x = 3.1416$ is equal to 8.0972.

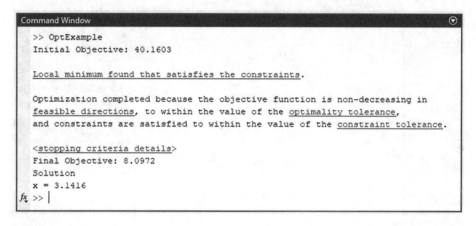

Fig. 11.14 Output of the code in Fig. 11.13

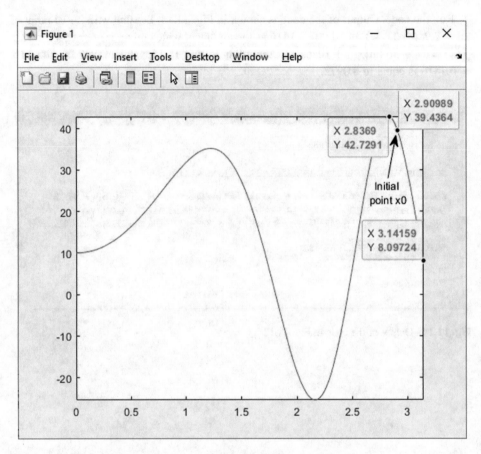

Fig. 11.15 Selected initial point and optimum points in its neighborhood

11.3   Global Minimum

In the previous section you learned how to obtain the local minimum. In this section you will learn how to find the global minimum. Let's take a look to the graph of the function studied in the previous section. The global minimum of this function (in the $[0, \pi]$ interval) occurs around the x = 2.16 (Fig. 11.16).

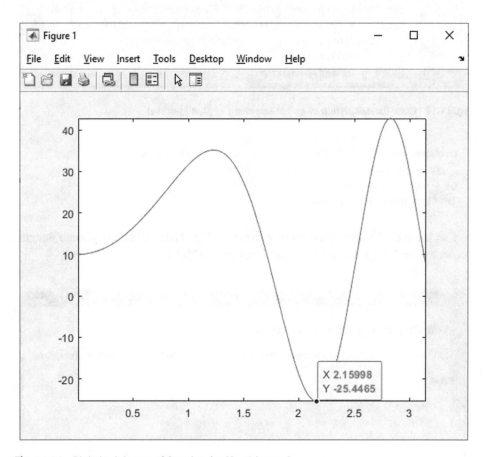

Fig. 11.16   Global minimum of function for $[0, \pi]$ interval

Let's use the MATLAB to find the global minimum of $f(x) = x^3 + 30\sin(x^2) + 10\cos(x)$ for $[0, \pi]$ interval. Enter the following code to the MATLAB editor (Fig. 11.17).

```
clc
fun= @(x) x^3+30*sin(x^2)+10*cos(x);
rng default % For reproducibility
opts = optimoptions(@fmincon,'Algorithm','sqp');
```

```
Editor - untitled*                                                    ⊙
  untitled      ×    +
  1      clc
  2      fun= @(x) x^3+30*sin(x^2)+10*cos(x);
  3
  4      rng default % For reproducibility
  5      opts = optimoptions(@fmincon,'Algorithm','sqp');
  6      problem = createOptimProblem('fmincon','objective',...
  7          fun,'x0',1,'lb',0,'ub',pi,'options',opts);
  8      gs = GlobalSearch;
  9      [x,f] = run(gs,problem)
```

Fig. 11.17 Code for calculation of global minimum in $[0, \pi]$ interval

```
problem = createOptimProblem('fmincon','objective',...
    fun,'x0',1,'lb',0,'ub',pi,'options',opts);
gs = GlobalSearch;
[x,f] = run(gs,problem)
```

Run the code. Output of the code is shown in Fig. 11.18. Obtained global minimum occurs at $x = 2.1607$ and value of cost function is -25.4466 there.

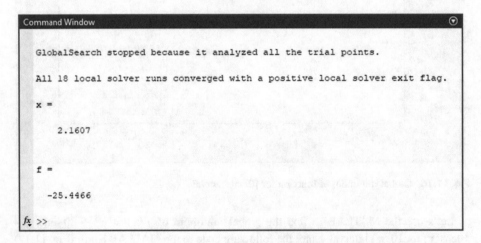

```
Command Window                                                           ⊙

  GlobalSearch stopped because it analyzed all the trial points.

  All 18 local solver runs converged with a positive local solver exit flag.

  x =

      2.1607


  f =

    -25.4466

fx >>
```

Fig. 11.18 Output of the code in Fig. 11.17

11.4   fminsearch Function

In the previous section we used the fmincon function to solve our optimization problem. fmincon can solve constrained optimizations. When your problem is unconstrained, you can fill the constraints with null matrix (i.e., []), or you can use fminsearch function which is designed to solve unconstrained optimizations.

Let's use the fminsearch function to minimize the $f(x) = x^3 + 30\sin(x^2) + 10\cos(x)$ subject to no constraints. Assume that initial point is x0 = 1. The code shown in Fig. 11.19 do this job for us. Output of this code is shown in Fig. 11.20.

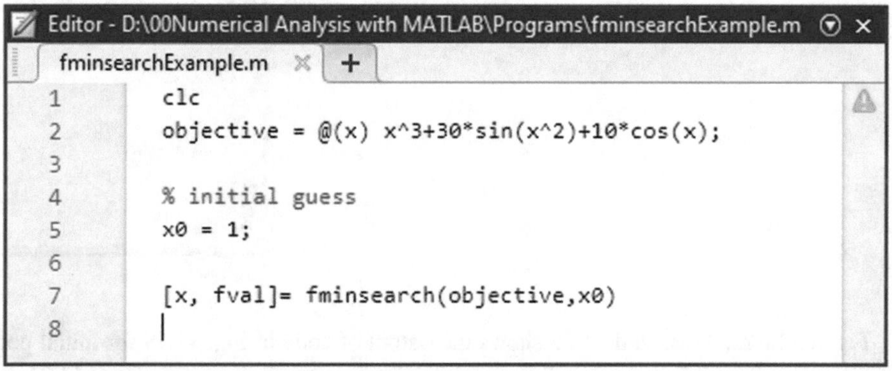

```
Editor - D:\00Numerical Analysis with MATLAB\Programs\fminsearchExample.m    ⊙ ✕

fminsearchExample.m    ✕    +

1        clc
2        objective = @(x) x^3+30*sin(x^2)+10*cos(x);
3
4        % initial guess
5        x0 = 1;
6
7        [x, fval]= fminsearch(objective,x0)
8        |
```

Fig. 11.19 Unconstrained minimization with fminsearch function. Initial point is x0 = 1

Fig. 11.20 Output of the code
in Fig. 11.19 with initial point
x0 = 1

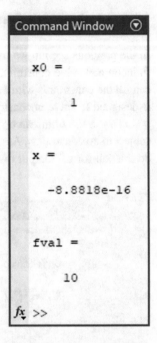

Figures 11.21, 11.22 and 11.23 shows the output of code in Fig. 11.19 for initial point
of x0 = 2, x0 = 2.5 and x0 = 2.9, respectively. The result shown in Figs. 11.21 and
11.22 are the same as results shown in Figs. 11.8 and 11.10. However, the result obtained
in Fig. 11.23 is not the same as the result in Fig. 11.14. In order to see why, remember
that in Fig. 11.14 minimization is constrained $(0 < x < \pi)$ but in Fig. 11.23, we have no
constraint. Graph of $f(x) = x^3 + 30 \sin(x^2) + 10 \cos(x)$ for $[0, 1.2\pi]$ interval is drawn
with the aid of the code shown in Fig. 11.24. Output of the code is shown in Fig. 11.25.
Note that the initial point x0 = 2.9 lies between two optimum points: 2.822 and 3.299.
Since value of function is smaller at point x = 3.299, MATLAB returns value of function
at x = 3.299 in Fig. 11.25.

Fig. 11.21 Output of the code in Fig. 11.19 with initial point x0 = 2

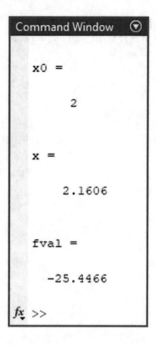

Fig. 11.22 Output of the code in Fig. 11.19 with initial point x0 = 2.5

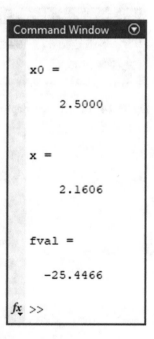

Fig. 11.23 Output of the code
in Fig. 11.19 with initial point
x0 = 2.9

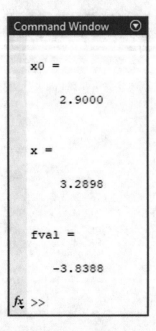

Fig. 11.24 Drawing the graph of $f(x) = x^3 + 30 \sin\left(x^2\right) + 10 \cos(x)$ for $[0, 1.2\pi]$ interval

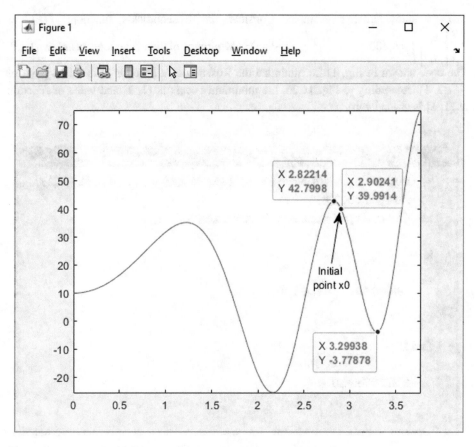

Fig. 11.25 Selected initial point and optimum points in its neighborhood

Let's study another example. Consider the Rosenbrock's function: $f(x) =$ $f\left(\begin{bmatrix} x_1 \\ x_2 \end{bmatrix}\right) = 100(x_2 - x_1^2)^2 + (1 - x_1)^2$. Minimum of this function occurs at $[1, 1]$. The code shown in Fig. 11.26 minimize the Rosenbrock's function with initial point of $[-1.2, 1]$. According to Fig. 11.26, the minimum occurs at $[1, 1]$ and value of function at $[1, 1]$ is around zero.

```
Command Window                                                      ⊙

  >> fun = @(x)100*(x(2) - x(1)^2)^2 + (1 - x(1))^2;
  >> x0=[-1.2,1];
  >> [x,fval]=fminsearch(fun,x0)


  x =


        1.0000      1.0000



  fval =


        8.1777e-10

fx >> |
```

Fig. 11.26 Minimization of Rosenbrock function x0 = [−1.2, 1]

11.5 Constrained Optimization

The fmincon command finds a constrained minimum of a function of several variables. fmincon attempts to solve problems of the form:

$$\min_{x} f(x) \quad \text{subject to:} \begin{cases} Ax \leq b, \quad A_{eq}x = b_{eq} \quad (linear\ constraints) \\ C(x) < 0, \quad C_{eq}(x) = 0\ (nonlinear\ constraints) \\ LB \leq x \leq UB \qquad\qquad (bounds) \end{cases} \quad (11.2)$$

Let's study an example:

$$\min x_1 x_4(x_1 + x_2 + x_3) + x_3$$

subject to: $x_1x_2x_3x_4 \geq 25$

$$x_1^2 + x_2^2 + x_3^2 + x_4^2 = 40$$

$$1 \leq x_1, x_2, x_3, x_4 \leq 5$$

$$x_0 = (1, 5, 5, 1) \tag{11.3}$$

Open the MATLAB editor and write the nlcon function (Fig. 11.27).

```
Editor - D:\nlcon.m                                    ⊙ ✕

  nlcon.m   ✕   +

1     % create file nlcon.m for nonlinear constraints
2     function [c,ceq] = nlcon(x)
3         c = 25.0 - x(1)*x(2)*x(3)*x(4);
4         ceq = sum(x.^2) - 40;


Command Window                                          ⊙

fx >>
```

Fig. 11.27 Entering the non-linear constraints into a function

Write the code shown in Fig. 11.28 in order to solve the problem. Output of this code is shown in Fig. 11.28. According to Fig. 11.29, the local minimum obtained for initial point of [1, 5, 5, 1] is [1, 4.743, 3.8212, 1.3794]. Value of the function at the obtained point is 17.014.

```
Editor - D:\optimizationProblemSolver.m                                    ⊙  ×
  nlcon.m  ×   optimizationProblemSolver.m  ×  +
  1       objective = @(x) x(1)*x(4)*(x(1)+x(2)+x(3))+x(3);              ✓
  2
  3       % initial guess
  4       x0 = [1,5,5,1];
  5
  6       % variable bounds
  7       lb = 1.0 * ones(4);
  8       ub = 5.0 * ones(4);
  9
  10      % show initial objective
  11      disp(['Initial Objective: ' num2str(objective(x0))])
  12
  13      % linear constraints
  14      A = [];
  15      b = [];
  16      Aeq = [];
  17      beq = [];
  18
  19      % nonlinear constraints
  20      nonlincon = @nlcon;
  21
  22  ⊟   % optimize with fmincon
  23      %[X,FVAL,EXITFLAG,OUTPUT,LAMBDA,GRAD,HESSIAN]
  24      % = fmincon(FUN,X0,A,B,Aeq,Beq,LB,UB,NONLCON,OPTIONS)
  25      x = fmincon(objective,x0,A,b,Aeq,beq,lb,ub,nonlincon);
  26
  27      % show final objective
  28      disp(['Final Objective: ' num2str(objective(x))])
  29
  30      % print solution
  31      disp('Solution')
  32      disp(['x1 = ' num2str(x(1))])
  33      disp(['x2 = ' num2str(x(2))])
  34      disp(['x3 = ' num2str(x(3))])
  35      disp(['x4 = ' num2str(x(4))])
```

Fig. 11.28 Code for solving the given constrained minimization

```
Command Window                                                                    ⊙
   >> optimizationProblemSolver
   Initial Objective: 16
   Warning: Length of lower bounds is > length(x); ignoring extra bounds.
   > In checkbounds (line 27)
   In fmincon (line 339)
   In optimizationProblemSolver (line 25)
   Warning: Length of upper bounds is > length(x); ignoring extra bounds.
   > In checkbounds (line 41)
   In fmincon (line 339)
   In optimizationProblemSolver (line 25)

   Local minimum found that satisfies the constraints.

   Optimization completed because the objective function is non-decreasing in
   feasible directions, to within the value of the optimality tolerance,
   and constraints are satisfied to within the value of the constraint tolerance.

   <stopping criteria details>
   Final Objective: 17.014
   Solution
   x1 = 1
   x2 = 4.743
   x3 = 3.8212
   x4 = 1.3794
fx >> |
```

Fig. 11.29   Output of the code in Fig. 11.28

Now let's find the global minimum of the obtained function. The code shown below do this for us.

```
clc
fun= @(x) x(1)*x(4)*(x(1)+x(2)+x(3))+x(3);
rng default % For reproducibility
opts = optimoptions(@fmincon,'Algorithm','sqp');
problem = createOptimProblem('fmincon','objective',...
    fun,'x0',[1 5 5 1],'lb',[1 1 1 1],'ub',[5 5 5 5],'options',opts);
gs = GlobalSearch;
[x,f] = run(gs,problem)
```

Enter the code to MATLAB editor (Fig. 11.30).

```
Editor - D:\00Numerical Analysis with MATLAB\Programs\OptimGlobal.m
  OptimGlobal.m  ×  +
1        clc
2        fun= @(x) x(1)*x(4)*(x(1)+x(2)+x(3))+x(3);
3
4        rng default % For reproducibility
5        opts = optimoptions(@fmincon,'Algorithm','sqp');
6        problem = createOptimProblem('fmincon','objective',...
7            fun,'x0',[1 5 5 1],'lb',[1 1 1 1],'ub',[5 5 5 5],'options',opts);
8        gs = GlobalSearch;
9        [x,f] = run(gs,problem)
```

Fig. 11.30 Code for calculation of global minimum

Run the code. According to Fig. 11.31, the global minimum occurs at [1, 1, 1, 1] and value of cost function at this point is 4 (Fig. 11.32).

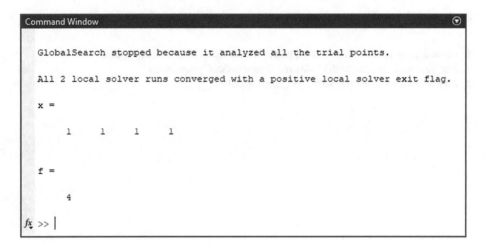

Fig. 11.31 Output of code in Fig. 11.30

Fig. 11.32 Schematic for
exercise 1

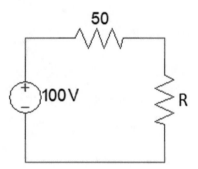

Exercises

1. Select the value of resistor R such that power dissipated in it has maximum value. **Hint**:
   You need to maximize $P = RI^2 = R\left(\frac{100}{50+R}\right)^2$. In other words you need to minimize
   $f(R) = -R\left(\frac{100}{50+R}\right)^2$.
2. fminbnd function can be used for minimization of single variable bounded nonlinear
   functions.
   (a) See the help of this function in order to see how it works.
   (b) Use this function to find the minimum of $f(x) = x^3 + 30\sin(x^2) + 10\cos(x)$ for
       $[0, \pi]$ interval.
3. Use MATLAB to solve max $x.y$ subject to $0 \le x \le 10, 0 \le y \le 10$ and $x + y = 10$.

References for Further Study

1. Optimization Toolbox User Guide, https://www.mathworks.com/help/optim/
2. http://apmonitor.com/che263/index.php/Main/MatlabOptimization
3. https://www.youtube.com/watch?v=Q2zgz0ag0L0

Printed in the United States
by Baker & Taylor Publisher Services